Essential Mathematics for Life

BOOK 3

Percents and Proportions

Fourth
Edition

GLENCOE
McGraw-Hill

New York, New York
Columbus, Ohio
Woodland Hills, California
Peoria, Illinois

Authors

Mary S. Charuhas
Associate Dean
College of Lake County
Grayslake, Illinois

Dorothy McMurty
District Director of ABE, GED,
 ESL
City Colleges of Chicago
Chicago, Illinois

The Mathematics Faculty
American Preparatory Institute
Killeen, Texas

Contributing Writers

Kathryn S. Harr
Mathematics Instructor
Pickerington, Ohio

Priscilla Ware
Educational Consultant and
 Instructor
Columbus, Ohio

Dr. Pearl Chase
Professional Consultants of Dallas
Cedar Hill, Texas

Previous Edition's
Academic Editors

Dr. Violet M. Malone
Past President of Adult
 Education of the USA
 (now AAACE)

Dale R. Jordan

Contributing Editors and Reviewers

Barbara Warner
Monroe Community College
Rochester, New York

Anita Armfield
York Technical College
Rock Hill, South Carolina

Judy D. Cole
Lafayette Regional Technical
 Institute
Lafayette, Louisiana

Mary Fincher
New Orleans Job Corps
New Orleans, Louisiana

Cheryl Gunderson
Rusk Community Learning
 Center
Ladysmith, Wisconsin

Cynthia A. Love
Columbus City Schools
Columbus, Ohio

Joyce Claar
South Westchester BOCES
Valhalla, New York

John Grabowski
St. Joseph Hill Academy
Staten Island, New York

Virginia Victor
Maple Run Youth Center
Cumberland, Maryland

Sandi Braga
College of South Idaho
Twin Falls, Idaho

Maggie Cunningham
Adult Education
Schertz, Texas

Sylvia Gilliard
Naval Consolidated Brig
Charleston, South Carolina

Eva Eaton-Smith
Cecil Community College
Elkton, Maryland

Fabienne West
John C. Calhoun State
 Community College
Decatur, Alabama

Photo credits: Cover, © Ralph Mercer/Tony Stone Images; 4, Cobalt Productions; 8, Larry Hamill; 15, Alan Carey; 30, 31, Rick Weber; 31, Ken Frick; 31, Rick Weber/Courtesy The Ohio Theatre, owned and operated by CAPA; 53, Aaron Haupt/Courtesy of Ethan Allen; 72, Mary Lou Uttermohlen; 84, 91, Aaron Haupt; 93, 102, Aaron Haupt/Courtesy of Ethan Allen; 132, Aaron Haupt/Courtesy T. Marzetti Co.; 135, Doug Martin; 136, Elaine Comer; 138, Aaron Haupt; 150, Mark Burnett; 174, Ted Rice; 181, Glencoe file; 197, Matt Meadows; 202, Doug Martin; 215, Mak-I; 216, 219, Glencoe file; 220, Rick Weber.

Send all inquiries to:
Glencoe/McGraw-Hill
21600 Oxnard St., Suite 500
Woodland Hills, CA 91367-4906

ISBN: 0-02-802610-1

7 8 9 10 11 12   066   04 03 02 01

C O N T E N T S

Percents

Unit 1 Review of Decimals and Fractions

Unit 2 Percents

Unit 3 Solving Percent Problems

Unit 4 Percent Word Problems

Unit 5 Using Percents

Unit 6 Ratios and Proportions

Review of Decimals and Fractions

Add or subtract.

1. $15 - 4.23 =$ _____

2. $124.09 + 4.12 + 67.43 =$ _____

3. $43.54 - 0.56 =$ _____

4. $900.1 + 7.89 + 40.7 =$ _____

5. $13.79 - 0.90 =$ _____

6. $12.4 + 9,004.7 + 6.98 =$ _____

Multiply.

7.
$$\begin{array}{r} 89 \\ \times\ .09 \\ \hline \end{array}$$

8.
$$\begin{array}{r} 0.3 \\ \times\ 0.2 \\ \hline \end{array}$$

9.
$$\begin{array}{r} 3.478 \\ \times\ \ \ \ 64 \\ \hline \end{array}$$

10.
$$\begin{array}{r} 78.25 \\ \times\ \ \ .03 \\ \hline \end{array}$$

11.
$$\begin{array}{r} 958 \\ \times\ 1.34 \\ \hline \end{array}$$

12.
$$\begin{array}{r} 1,370 \\ \times\ .026 \\ \hline \end{array}$$

Divide.

13. $0.6\overline{)360}$ **14.** $0.25\overline{)5}$ **15.** $0.004\overline{)2.6}$

Put in order from the smallest to the largest.

16.   0.04   0.4 $\frac{1}{4}$ **17.** $\frac{7}{8}$ 0.08 $\frac{1}{3}$

_____   _____

Add or subtract.

18. $2\frac{3}{5}$ **19.** 8 **20.** $4\frac{1}{3}$
   $+\,3\frac{1}{5}$ $-\,\frac{5}{7}$ $-\,\frac{3}{8}$

21. $5\frac{1}{4}$ **22.** $6\frac{4}{9}$ **23.** $7\frac{1}{7}$
   $+\,6\frac{2}{4}$ $-\,3\frac{2}{9}$ $+\,3\frac{2}{7}$

24. 10 **25.** $3\frac{1}{5}$ **26.** 16
   $-\,\frac{1}{8}$ $-\,\frac{1}{10}$ $+\,9\frac{1}{4}$

Multiply or divide.

27. $40 \times 1\frac{1}{8} =$ **28.** $8\frac{1}{4} \div \frac{11}{20} =$ **29.** $\frac{5}{6} \times 36 =$

_____   _____   _____

Adding and Subtracting Decimals

When adding or subtracting decimals, follow these steps:

Step 1 Write the numbers in columns. Line up decimal points.

Step 2 Remember whole numbers have a decimal point at the right.

Step 3 Add zeros to the right of the decimal point if needed.

Step 4 Add or subtract as with whole numbers.

Step 5 Bring the decimal point straight down into the answer.

Examples

A. Add.

$4 + 4.62 =$

```
  4.00
+ 4.62
------
  8.62
```

B. Add.

$0.08 + 0.93 =$

```
  .08
+ .93
-----
 1.01
```

C. Subtract.

$6 - 2.312 =$

```
  6.000
- 2.312
-------
  3.688
```

D. Subtract.

$12.09 - 0.003 =$

```
 12.090
-   .003
--------
 12.087
```

Practice

Add or subtract.

1. $10 - 1.37 =$ _____

2. $454.008 + 5.12 + 0.063 =$ _____

3. $2.44 - 0.76 =$ _____

4. $1,001.1 + 1.9 + 20.05 =$ _____

5. $11.28 - 0.12 =$ _____

6. $565 + 0.25 + 1.78 =$ _____

7. $1,044.3 - 257.07 =$ _____

8. $5.54 + 10 + 1.003 =$ _____

9. $0.99 - 0.9 =$ _____

10. $956 + 77 + 0.88 =$ _____

Solve the following problems.

11. The sum of the speeds for the three fastest cars in the race was 15.705 minutes. If the second place winner's time was 5.305 minutes and the third place winner's time was 5.4 minutes, what was the time of the first place winner?
_____

12. Duncan earned $14.25 in tips on Friday and $24.88 in tips on Saturday. If he earned a total of $56.87 in tips on his weekend job, what did he earn in tips on Sunday?
_____

13. Lucy prepared a deposit slip for her checking account. She listed three items: her paycheck of $651.59, a travel claim of $198.28, and a store refund worth $43.72. If she deposited $849.59, how much did she get back in cash?
_____

Multiplying Decimals

When multiplying decimals, follow these steps:

Step 1 Multiply the numbers as if they were whole numbers.

Step 2 Count the number of decimal places (numbers to the **right** of the decimal) in both numbers. Add to find the total.

Step 3 Use the total to count off decimal places in the answer. Count from right to left.

Step 4 Fill in with zeros if more decimal places are needed.

Examples

Multiply.

A.
```
    3 0        0  place
  × 2 . 3    + 1  place
  ─────────   ─────────
  6 9 . 0      1  place
```

B.
```
    1 5        0  place
  × . 1 5    + 2  places
  ─────────
      7 5
    1 5
  ─────────
  2 . 2 5      2  places
```

C.
```
    . 0 0 5      3  places
  ×   . 0 7    + 2  places
  ─────────────  ─────────
  . 0 0 0 3 5    5  places
```

D.
```
    1 . 0 7 8      3  places
  ×     . 0 0 3  + 3  places
  ───────────────  ─────────
  . 0 0 3 2 3 4    6  places
```

Practice

Multiply.

1.
```
  4 5 3
× . 0 3
```

2.
```
  5 , 7 5 1
×   . 0 0 6
```

3.
```
  0 . 4 7
×     . 2
```

4.
```
    . 0 2 5
×       2 2
```

5.
```
  2 3 . 5 4
×       . 0 5
```

6.
```
  4 5 . 6 7
×       . 2 3
```

7.
```
    5 5 5
× 1 . 7 6
```

8.
```
    9 5 0
× . 0 8 7
```

5

9.	10.	11.	12.
11.1 × .203	2,591 × .8902	178.6 × .905	1.155 × .8321

Problem Solving

Solve the following problems.

13. The Chikaris family went out for dinner. There were three adults and two children. An adult dinner was priced at $10.55 each. Each child's dinner cost $4.95. What was the total cost of the meals, not including tax and tip? _____

14. The flower market had a sale. Roses were two for $6.99. Peonies sold for $1.99 each. Geraniums were $1.25 each. Nina bought six roses, two peonies, and five geraniums. What was her bill for the flowers? _____

15. Lee earns $14.24 an hour. He earns $21.36 per hour in overtime pay. What is his total weekly salary if he works 40 hours at his regular pay and 5 hours overtime? _____

Dividing Decimals

Examples

When dividing decimals by **whole numbers,** follow these steps:

Step 1 Divide as usual.

Step 2 Bring the decimal point straight up into the answer.

A.

$$
\begin{array}{r}
1.3 \\
3\overline{)3.9} \\
\underline{3} \\
9 \\
\underline{9}
\end{array}
$$

B.

$$
\begin{array}{r}
.25 \\
8\overline{)2.00} \\
\underline{1\,6} \\
40 \\
\underline{40}
\end{array}
$$

When dividing decimals or whole numbers by **decimals,** follow these steps:

Step 1 Move the decimal point in the divisor all the way to the right.

Step 2 Move the decimal point in the dividend the same number of decimal places to the right.

Step 3 Fill in empty decimal places with zeros.

Step 4 Bring the decimal point straight up into the answer.

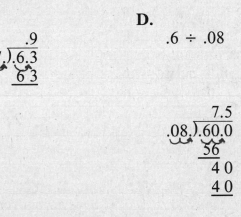

C.

$$
\begin{array}{r}
.9 \\
.7\overline{)6.3} \\
\underline{6\,3}
\end{array}
$$

D.

$$.6 \div .08$$

$$
\begin{array}{r}
7.5 \\
.08\overline{)60.0} \\
\underline{56} \\
4\,0 \\
\underline{4\,0}
\end{array}
$$

> **MATH HINT**
>
> **R**emember, in a whole number, the decimal point is on the right.

Practice

Divide.

1. 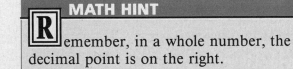 $.5\overline{)300}$

2. $.25\overline{)7}$

3. $.003\overline{)2.1}$

4. $8\overline{)64.8}$ **5.** $1.6\overline{).72}$ **6.** $63.9 \div 0.009$ _____

7. $1.6 \div 0.04$ _____   **8.** $.25 \div 0.005$ _____   **9.** $5.6 \div 0.08$ _____

Problem Solving

Solve the following problems.

10. The city wants various community groups to sponsor 1.5 miles of highway each. There are 30 miles of highway that need to be sponsored. How many community groups will be needed?

_____

11. The telephone company needs 1,250 feet of wire to install new lines. If each spool of wire holds 12.5 feet of wire, how many spools will be used?

_____

12. The Kenya family inherited $13,600 to be shared equally among four brothers. How much did each brother receive?

_____

Fractions

Fractions are used in special mathematics operations, as shown in the examples below.

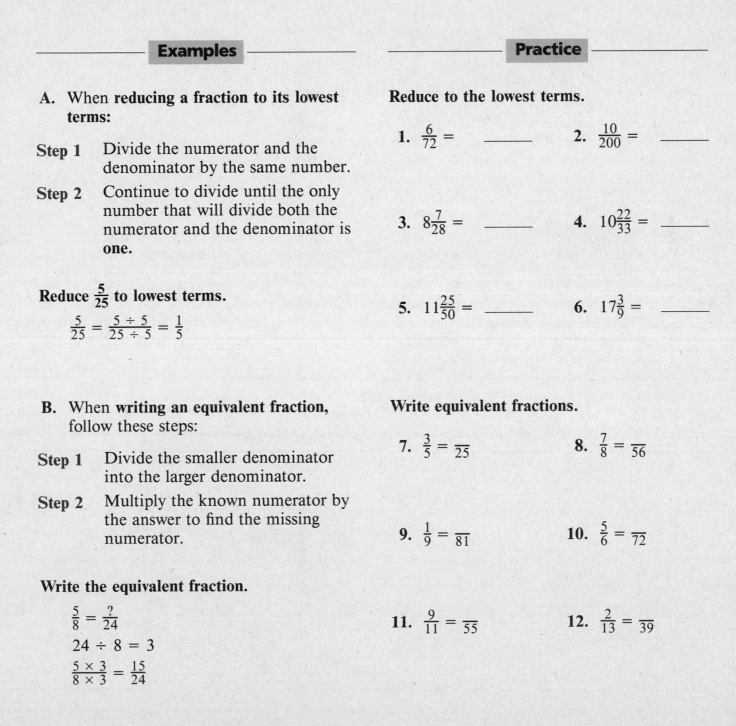

——————— **Examples** ———————

A. When **reducing a fraction to its lowest terms:**

Step 1 Divide the numerator and the denominator by the same number.

Step 2 Continue to divide until the only number that will divide both the numerator and the denominator is **one.**

Reduce $\frac{5}{25}$ to lowest terms.

$$\frac{5}{25} = \frac{5 \div 5}{25 \div 5} = \frac{1}{5}$$

B. When **writing an equivalent fraction,** follow these steps:

Step 1 Divide the smaller denominator into the larger denominator.

Step 2 Multiply the known numerator by the answer to find the missing numerator.

Write the equivalent fraction.

$$\frac{5}{8} = \frac{?}{24}$$

$$24 \div 8 = 3$$

$$\frac{5 \times 3}{8 \times 3} = \frac{15}{24}$$

——————— **Practice** ———————

Reduce to the lowest terms.

1. $\frac{6}{72} =$ _____

2. $\frac{10}{200} =$ _____

3. $8\frac{7}{28} =$ _____

4. $10\frac{22}{33} =$ _____

5. $11\frac{25}{50} =$ _____

6. $17\frac{3}{9} =$ _____

Write equivalent fractions.

7. $\frac{3}{5} = \frac{}{25}$

8. $\frac{7}{8} = \frac{}{56}$

9. $\frac{1}{9} = \frac{}{81}$

10. $\frac{5}{6} = \frac{}{72}$

11. $\frac{9}{11} = \frac{}{55}$

12. $\frac{2}{13} = \frac{}{39}$

C. When **finding a common denominator,** find a number all denominators will divide into evenly.

Find the common denominator.

$\frac{1}{8}$ $\frac{1}{24}$ $\frac{1}{12}$ 8 divides evenly into 24.
12 divides evenly into 24.
24 divides evenly into 24.

24 is the common denominator.

Find a common denominator.

13. $\frac{2}{3}$ $\frac{1}{9}$ _____

14. $\frac{1}{2}$ $\frac{1}{4}$ $\frac{1}{5}$ _____

15. $\frac{7}{8}$ $\frac{3}{16}$ _____

16. $\frac{2}{3}$ $\frac{3}{4}$ $\frac{4}{5}$ _____

17. $\frac{5}{9}$ $\frac{1}{2}$ _____

18. $\frac{11}{12}$ $\frac{1}{5}$ $\frac{2}{30}$ _____

D. When **renaming mixed numbers,** follow these steps:

Step 1 Rename the mixed number as an improper fraction. Use the same denominator for the improper fraction as in the given fractions.

Step 2 Add the fractions.

Rename the mixed number.

$$7\frac{1}{4} = 6 + 1 + \frac{1}{4}$$
$$= 6 + \frac{4}{4} + \frac{1}{4} \Big\} \quad \text{Step 1}$$
$$= 6\frac{5}{4} \qquad\qquad \text{Step 2}$$

Rename these mixed numbers.

19. $9\frac{1}{2} = 8\frac{}{2}$ **20.** $6 = 5\frac{}{3}$

21. $9 = 8\frac{}{12}$ **22.** $21\frac{7}{8} = 20\frac{}{8}$

23. $47\frac{3}{8} = 46\frac{}{8}$ **24.** $19\frac{1}{2} = 18\frac{}{2}$

E. When **changing an improper fraction to a mixed or whole number,** follow these steps:

Step 1 Divide the numerator by the denominator.

Step 2 If there is a remainder, write it as a fraction.

Change the fraction to a mixed number.

$$\frac{15}{7} = 7\overline{)15} \quad = 2\frac{1}{7}$$
$$\phantom{\frac{15}{7} = 7}\underline{14}$$
$$\phantom{\frac{15}{7} = 7)}1R$$

Change these to mixed or whole numbers.

25. $\frac{22}{5} =$ _____   **26.** $\frac{20}{20} =$ _____

27. $\frac{30}{4} =$ _____   **28.** $\frac{48}{9} =$ _____

29. $\frac{99}{11} =$ _____   **30.** $\frac{36}{36} =$ _____

F. When **changing mixed numbers to improper fractions,** follow these steps:

Step 1 Multiply the denominator by the whole number. Add the numerator.

Step 2 Write the result over the original denominator.

Change the mixed number to an improper fraction.

$3\frac{1}{4}$ ← numerator
← denominator

$4 \times 3 + 1 = 13$

$3\frac{1}{4} = \frac{13}{4}$

Change these to improper fractions.

31. $5\frac{2}{5} =$ _____   **32.** $9\frac{1}{4} =$ _____

33. $33\frac{1}{3} =$ _____   **34.** $16\frac{3}{4} =$ _____

35. $7\frac{1}{2} =$ _____   **36.** $8\frac{1}{3} =$ _____

Multiplying Fractions

To multiply fractions, follow these steps:

Step 1 Change all mixed numbers to improper fractions.

Step 2 Cancel if possible.

Step 3 Multiply straight across.

Step 4 Reduce.

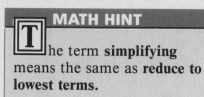

MATH HINT

The term **simplifying** means the same as **reduce to lowest terms.**

Examples

Multiply.

A. $\frac{3}{4} \times \frac{1}{2} = \frac{3}{8}$ Multiply straight across.

B. $\frac{\overset{1}{\cancel{3}}}{\cancel{7}} \times \frac{\overset{1}{\cancel{7}}}{\cancel{9}} = \frac{1}{3}$ Cancel.
Multiply straight across.

C. $1\frac{1}{3} \times 4\frac{5}{8}$

$= \frac{4}{3} \times \frac{37}{8}$ Change mixed numbers to improper fractions.

$= \frac{\overset{1}{\cancel{4}}}{3} \times \frac{37}{\underset{2}{\cancel{8}}}$ Cancel.

$= \frac{37}{6}$ Multiply straight across.

$= 6\frac{1}{6}$ Reduce.

Practice

Multiply.

1. $\frac{1}{2} \times \frac{1}{3} =$ _____

2. $1\frac{1}{2} \times \frac{8}{9} =$ _____

3. $\frac{2}{5} \times \frac{5}{12} \times \frac{8}{13} =$ _____

4. $25 \times \frac{1}{8} =$ _____

5. $8\frac{1}{3} \times 4\frac{3}{5} \times \frac{15}{46} =$ _____

6. $\frac{5}{8} \times \frac{7}{25} =$ _____

7. $5\frac{2}{5} \times 3\frac{1}{3} =$ _____

8. $\frac{5}{8} \times 4 \times \frac{16}{25} =$ _____

Solve the following problems.

9. The Hans family wanted to put baseboard around their living room. Each wall was $12\frac{1}{2}$ feet long. How much lumber would they need for all four walls?

_____

10. Suzi made dresses for a wedding. Each dress required $2\frac{2}{3}$ yards of material. How much material would be needed for 8 dresses?

_____

11. Trace built a bookcase. Each shelf was $4\frac{1}{2}$ feet long. How much lumber will he need for 5 shelves?

_____

12. Cassie bought ribbon to make bows. If each bow needed $6\frac{1}{2}$ inches of ribbon, how much ribbon will she need for 7 bows?

_____

Dividing by Fractions

When dividing by fractions, follow these steps:

Step 1 Change all mixed or whole numbers to improper fractions.

Step 2 Invert the divisor.

Step 3 Cancel if possible.

Step 4 Multiply straight across.

Step 5 Reduce.

> **MATH HINT**
>
> To invert means the numerator becomes the denominator, and the denominator becomes the numerator.

Examples

Divide.

A. $\frac{1}{2} \div \frac{3}{8} = \frac{1}{2} \times \frac{\overset{4}{8}}{3} = \frac{4}{3} = 1\frac{1}{3}$

B. $1\frac{1}{3} \div 2\frac{5}{6} = \frac{4}{3} \div \frac{17}{6} = \frac{4}{3} \times \frac{\overset{2}{6}}{17} = \frac{8}{17}$

C. $2\frac{2}{3} \div 4 = \frac{8}{3} \div \frac{4}{1} = \frac{\overset{2}{8}}{3} \times \frac{1}{\underset{1}{4}} = \frac{2}{3}$

Practice

Divide. Reduce to lowest terms.

1. $\frac{1}{8} \div \frac{1}{8} =$ _____

2. $\frac{2}{3} \div \frac{4}{9} =$ _____

3. $\frac{11}{12} \div \frac{3}{4} =$ _____

4. $22 \div \frac{1}{2} =$ _____

5. $15 \div \frac{3}{4} =$ _____

6. $2\frac{3}{8} \div 19 =$ _____

7. $6\frac{3}{4} \div 9 =$ _____

8. $\frac{2}{5} \div 1\frac{3}{5} =$ _____

9. $13\frac{3}{4} \div 5 =$ _____

10. $\frac{7}{8} \div 1\frac{5}{16} =$ _____ **11.** $5\frac{2}{9} \div \frac{1}{27} =$ _____ **12.** $1\frac{1}{2} \div \frac{3}{4} =$ _____

Problem Solving

Solve the following problems.

13. An apartment building has a circular garden surrounded by a sidewalk. The owners want to plant flowers every $\frac{1}{2}$ foot along the 80-foot edge. How many flowers will they need to plant?

_____

14. Jackson needs $\frac{1}{4}$ foot of lumber to make a garden marker. If he has wood that measures 10 feet, how many markers can he make?

_____

15. If a panel that measures $3\frac{1}{2}$ feet wide is cut into three equal sections, what is the width of each section?

_____

16. Gordon has $18\frac{1}{2}$ feet of cable line. If he has to put an equal amount of cable in three houses, how much cable will he use for each house?

_____

Adding and Subtracting Fractions

When adding and subtracting fractions, follow these steps:

Step 1 Find a common denominator for the fractions. Change both fractions to equivalent fractions with common denominators.

Step 2 If the top fraction in a subtraction problem is less than the bottom fraction, rename the top whole number.

Step 3 Add or subtract the numerators of the fractions. Add or subtract the whole numbers.

Step 4 Reduce all answers.

--- **Examples** ---

A. Subtract.

$$1\tfrac{3}{5}$$
$$-\ \tfrac{2}{5}$$
$$1\tfrac{1}{5}$$

B. Add.

$$4\tfrac{2}{3} = 4\tfrac{8}{12}$$
$$+\ 3\tfrac{1}{4} = 3\tfrac{3}{12}$$
$$7\tfrac{11}{12}$$

C. Add.

$$5\tfrac{3}{10} = 5\tfrac{3}{10}$$
$$6\tfrac{4}{5} = 6\tfrac{8}{10}$$
$$+\ 1\tfrac{1}{2} = 1\tfrac{5}{10}$$
$$1\,2\tfrac{16}{10} = 13\tfrac{6}{10} = 13\tfrac{3}{5}$$

D. Subtract.

$$9$$
$$-\ \tfrac{2}{7}$$

Step 1
$\tfrac{2}{7}$ cannot be subtracted from a whole number.

Step 2
Rename the 7 so it is a mixed number. For the denominator of the new fraction, choose the denominator of $\tfrac{2}{7}$.

$$7 = 6 + \tfrac{7}{7}$$

$$\overset{8}{\cancel{9}}\,\tfrac{7}{7}$$
$$-\ \tfrac{2}{7}$$
$$8\tfrac{5}{7}$$

Step 3
Subtract the numerators of the fractions. Then subtract the whole numbers.

E. Subtract.

$$5 \frac{1}{3} = 5 \frac{8}{24}$$
$$- \quad \frac{3}{8} = \quad \frac{9}{24}$$

Step 1
Find a common denominator. 9 cannot be subtracted from 8.

Step 2
Rename the 5 as a mixed number, using the common denominator.

$$\overset{4}{\cancel{5}} \frac{8}{24} + \frac{24}{24}$$
$$- \quad \frac{9}{24}$$

$$5 = 4 + \frac{24}{24}$$

Add $4 + \frac{24}{24}$ to $\frac{8}{24}$.

$$4 + \frac{24}{24} + \frac{8}{24} = 4 + \frac{32}{24}$$

$$4 \frac{32}{24}$$
$$- \quad \frac{9}{24}$$
$$4 \frac{23}{24}$$

Step 3
Now subtract.

Practice

Add or subtract.

1.
$$1 \frac{3}{10}$$
$$+ 2 \frac{3}{10}$$

2.
$$1 2$$
$$- \quad \frac{2}{3}$$

3.
$$6$$
$$- \quad \frac{5}{8}$$

4.
$$1 7 \frac{23}{40}$$
$$- \quad 5 \frac{11}{40}$$

5.
$$2 \frac{4}{9}$$
$$+ \quad \frac{5}{9}$$

6.
$$1 4 \frac{4}{7}$$
$$+ \quad 2 \frac{4}{5}$$

7.
$$1 5 \frac{1}{3}$$
$$- \quad 1 \frac{3}{4}$$

8.
$$7 \frac{7}{8}$$
$$+ \quad 6$$

9.
$$23 \frac{2}{11}$$
$$- \quad 1 \frac{4}{33}$$

17

10. $17\frac{8}{11} + 10\frac{2}{3} =$ _____ **11.** $9\frac{1}{2} + 7\frac{1}{2} =$ _____ **12.** $13\frac{2}{3} - \frac{2}{3} =$ _____

13. $10\frac{3}{4} - \frac{1}{3} =$ _____ **14.** $28\frac{8}{9} - 8\frac{7}{8} =$ _____ **15.** $13\frac{5}{11} + 5\frac{1}{3} =$ _____

16. $95\frac{3}{4} - 42\frac{5}{6} =$ _____ **17.** $6\frac{3}{5} + 1\frac{2}{7} =$ _____ **18.** $17\frac{23}{40} - 8\frac{11}{40} =$ _____

Problem Solving

Solve the following problems. Circle the correct answer.

19. Last week Cheung worked $9\frac{2}{3}$ hours overtime. This week he worked $7\frac{1}{6}$ hours overtime. How much overtime did Cheung work in all?

    (1) $17\frac{5}{6}$ hours   **(2)** $16\frac{5}{6}$ hours

    (3) $16\frac{1}{2}$ hours   **(4)** 17 hours

20. How many more hours did Cheung work last week than this week?

    (1) $2\frac{1}{2}$ hours   **(2)** $2\frac{1}{6}$ hours

    (3) $9\frac{2}{3}$ hours   **(4)** 2 hours

Writing Equal Decimals and Fractions

When writing fractions as decimals, follow these steps:

Step 1 The fraction bar indicates division. Divide the numerator by the denominator.

Step 2 Use a decimal point and zeros if needed.

Step 3 Carry out to 2 decimal places. Make any remainder a fraction.

Examples

Rewrite the fractions as decimals.

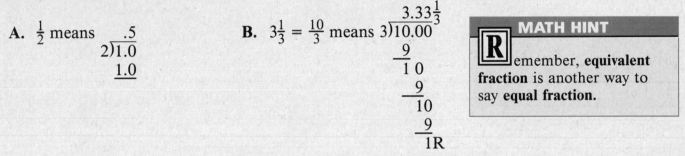

A. $\frac{1}{2}$ means

$$2\overline{)1.0} \quad .5$$
$$\underline{1.0}$$

B. $3\frac{1}{3} = \frac{10}{3}$ means

$$3\overline{)10.00} \quad 3.33\tfrac{1}{3}$$
$$\underline{9}$$
$$1\ 0$$
$$\underline{9}$$
$$10$$
$$\underline{9}$$
$$1R$$

> **MATH HINT**
> **R**emember, **equivalent fraction** is another way to say **equal fraction**.

When writing decimals as fractions, follow these steps:

Step 1 Draw a line under the decimal.

Step 2 Put a zero under every number.

Step 3 Write 1 for the decimal point.

Step 4 Drop the decimal point and reduce.

Examples

Rewrite the decimals as fractions.

C. $.358 \rightarrow \frac{358}{1,000} = \frac{179}{500}$

D. $1.005 \rightarrow 1\frac{005}{1,000} \rightarrow 1\frac{5}{1,000} = 1\frac{1}{200}$

Write as decimals.

1. $\frac{1}{10}$ _____

2. $5\frac{3}{8}$ _____

3. $\frac{4}{9}$ _____

4. $\frac{7}{8}$ _____

5. $1\frac{1}{2}$ _____

6. $\frac{2}{3}$ _____

7. $\frac{4}{5}$ _____

8. $\frac{7}{16}$ _____

9. $\frac{1}{6}$ _____

10. $\frac{3}{4}$ _____

11. $\frac{1}{9}$ _____

12. $\frac{3}{10}$ _____

13. $\frac{2}{5}$ _____

14. $11\frac{9}{10}$ _____

15. $\frac{1}{3}$ _____

16. $\frac{5}{6}$ _____

Write as fractions.

17. .125 _____

18. .25 _____

19. .225 _____

20. .32 _____

21. 5.01 _____

22. .2 _____

23. 7.2 _____

24. .001 _____

25. .85 _____

26. .003 _____

27. .625 _____

28. 4.1 _____

29. .0012 _____

30. 6.125 _____

31. .503 _____

32. 5.09 _____

Comparing Decimals and Fractions

To compare decimals and fractions, follow the steps in the **fraction** or **decimal** method.

Fraction Method

Step 1 Write both numbers as fractions using a common denominator.

Step 2 Compare numerators.

MATH HINT

Placeholders are zeros placed at the end of the numbers so that they have the same number of digits.

Decimal Method

Step 1 Write both numbers as decimals.

Step 2 Line up the decimal points.

Step 3 Fill in zeros as placeholders.

Step 4 Compare numbers.

Examples

Determine which is larger, $\frac{1}{10}$ or .003.

A. Solve by using the fraction method.

$$\frac{1}{10} = \frac{100}{1,000}$$
$$.003 = \frac{3}{1,000}$$
$$\frac{100}{1000} > \frac{3}{1000}$$

$\frac{1}{10}$ is larger,

or $\frac{1}{10} > .003$

B. Solve by using the decimal method.

$$\frac{1}{10} = .1 = .100$$
$$.003 = .003$$
$$.100 > .003$$
$$\frac{1}{10} > .003$$

Practice

Which is larger?

1. $\frac{3}{4}$ or .5 _____

2. .3 or $\frac{1}{4}$ _____

3. $\frac{8}{9}$ or 1.1 _____

4. .85 or $\frac{5}{6}$ _____

5. $\frac{1}{3}$ or .4 _____

6. .75 or $\frac{2}{3}$ _____

Which is smaller?

7. $\frac{1}{10}$ or .09 _____

8. .009 or $\frac{3}{100}$ _____

9. $\frac{3}{4}$ or .49 _____

10. $2\frac{4}{5}$ or 3.01 _____

11. $\frac{1}{8}$ or .2 _____

12. $\frac{1}{2}$ or .60 _____

Put in order from the smallest to the largest.

13. .09   .9 $\frac{1}{5}$

_____   _____   _____

14. $\frac{3}{8}$ $\frac{5}{9}$ .6

_____   _____   _____

15. .23   .032   .32

_____   _____   _____

16. .125   .1025   .025

_____   _____   _____

17. $\frac{3}{4}$ $\frac{2}{4}$ $\frac{7}{8}$

_____   _____   _____

18. $\frac{1}{2}$ $\frac{1}{8}$ .4

_____   _____   _____

19. .5   1.6   .05

_____   _____   _____

20. .02   11.15   .2

_____   _____   _____

Problem Solving With Decimals and Fractions

Throughout this book and all the other books in the series, you will be asked to solve a variety of word problems. Each unit will have a lesson called **Problem Solving.** In these lessons, you will solve everyday problems that use a particular math concept or computation that was introduced in the unit.

Use the following steps to help you solve the problems. Read the list carefully and try to apply the methods when you are answering word problems.

Step 1 Read the problem and underline the key words. These words will usually relate to some mathematics reasoning computation.

Step 2 Make a plan to solve the problem. Ask yourself, Should I add, subtract, multiply, divide, round, or compare? You may have to do more than one of these operations for the same problem.

Step 3 Find the solution. Use your math knowledge to find your answer.

Step 4 Check the answer. Ask yourself, Is the answer reasonable? Did you find what you were asked for?

Some key words for addition and subtraction are listed for you.

Addition		Subtraction	
altogether	total	decreased by	how much less
both	together	diminished by	how much more
increase	in all	difference	remainder
sum			

Example

Mrs. Rivera wants to bake cookies and a cake for her grandchildren. The cookie recipe calls for $1\frac{1}{3}$ cups of flour and the cake recipe needs $2\frac{1}{3}$ cups of flour. How much flour does she need altogether?

Step 1 Determine how much flour is needed. The key word is **altogether.**

Step 2 The key word indicates which operation should occur—addition.

Step 3 Solve the problem.

$$1 \frac{1}{3}$$
$$+2 \frac{1}{3}$$
$$\overline{3 \frac{2}{3}}$$

Step 4 Check the answer. Does it make sense that $3\frac{2}{3}$ is the total of $1\frac{1}{3} + 2\frac{1}{3}$? Yes, the answer is reasonable because $1\frac{1}{3}$ is more than 1; $2\frac{1}{3}$ is more than 2; thus, $3\frac{2}{3}$ is more than 3.

Practice
---

1. Consuela has a part-time job at a nursery. This week, she worked $3\frac{1}{2}$ hours on Monday, $3\frac{1}{2}$ hours on Tuesday, 2 hours on Wednesday, $2\frac{1}{4}$ hours on Thursday, and $2\frac{1}{4}$ hours on Friday. How many hours did she work in all? _____

2. Linc is putting up shelves in his children's closet. He needs $4\frac{1}{8}$ feet of wood for each shelf. How much wood does he need, if he plans to put up three shelves? _____

3. Carla bought $5\frac{7}{8}$ yards of lace for a wedding dress she is making. She used $3\frac{1}{4}$ yards. How much lace does she have left? _____

4. Andrew earned $90 this week. He spent $15.89 on gas for his car and $19.45 on food. How much does he have left from his earnings? _____

5. Nina invited her mother out for dinner. Nina's meal cost $7.95, and her mother's meal was $9.25. How much did she spend altogether? _____

————————— **Posttest** —————————

Add or subtract.

1. $24 - 14.63 =$ _____

2. $214.89 + 7.21 + 76.47 =$ _____

3. $89.32 - 0.66 =$ _____

4. $600.2 + 6.84 + 30.9 =$ _____

5. $45.76 - 0.80 =$ _____

6. $41.8 + 5,040.8 + 9.98 =$ _____

Multiply.

7. $\begin{array}{r} 1\,9 \\ \times\,.0\,9 \\ \hline \end{array}$

8. $4\frac{1}{8} \times 6\frac{1}{8} =$

9. $\begin{array}{r} 0\,.\,4 \\ \times\,0\,.\,2 \\ \hline \end{array}$

10. $\begin{array}{r} 4\,5\,.\,7\,8 \\ \times\qquad.0\,4 \\ \hline \end{array}$

11. $\begin{array}{r} 5\,7\,8 \\ \times\,3\,.\,1\,4 \\ \hline \end{array}$

12. $\begin{array}{r} 3\,,\,2\,1\,9 \\ \times\qquad.0\,2\,6 \\ \hline \end{array}$

Divide.

13. $.4\overline{)1{,}608}$

14. $\frac{9}{27} \div 3$

15. $.007\overline{)4.3}$

Put in order from the smallest to the largest.

16. $\frac{1}{5}$ .50   .05

17. $\frac{9}{8}$ .09 $\frac{8}{9}$

_____

_____

Add or subtract.

18. $3\frac{2}{4}$
    $+ 6\frac{1}{4}$
    _____

19. 6
    $- \frac{2}{5}$
    _____

20. $14\frac{1}{6}$
    $- 2\frac{1}{12}$
    _____

Problem Solving

Solve the following problems.

21. Gavin had $115.56 in his checking account. He deposited a refund check of $45.79, his paycheck of $345.67, and a cash amount of $123.93. How much does he have in his account now?   _____

22. Nick is building a birdhouse. He needs $3\frac{4}{5}$ feet of lumber. If he purchases $6\frac{3}{4}$ feet, how much will he have left after he builds the birdhouse?   _____

23. Millie is going to buy a car that costs $12,000. She has saved $\frac{1}{3}$ of the cost. How much has she saved?   _____

    Millie's sister is going to lend her $\frac{1}{4}$ of the cost. How much will her sister lend her?   _____

    How much money does Millie still need to purchase the car?   _____

26

Percents

The figure below is divided into 100 parts.

Use the figure to answer questions 1–3.

1. What percent of the squares are shaded? _____

2. What percent of the squares are not shaded? _____

3. What percent of the box has squares? _____

Change to decimals.

4. 25% = _____   5. 245% = _____   6. 1% = _____

7. $66\frac{2}{3}\%$ = _____   8. 77% = _____   9. .08% = _____

Change to fractions.

10. 90% = _____   11. $33\frac{1}{3}\%$ = _____   12. 288% = _____

13. 68% = _____   14. $\frac{1}{2}\%$ = _____   15. 4% = _____

Change to percents.

16. $.33\frac{1}{3}$ = _____

17. $1\frac{3}{4}$ = _____

18. 42.1 = _____

19. 1 = _____

20. $.0082$ = _____

21. $\frac{1}{5}$ = _____

Write in order from the smallest to the largest.

22. $\frac{1}{4}$ $.75$ $\frac{4}{5}$ 30% _____ _____ _____ _____

23. $.33$ $\frac{2}{5}$ $\frac{1}{4}$ 15% _____ _____ _____ _____

24. $.04$ 18% $\frac{2}{4}$ 2.1 _____ _____ _____ _____

Write in order from largest to smallest.

25. $\frac{1}{3}$ 35% $\frac{2}{3}$ 30% _____ _____ _____ _____

26. $\frac{2}{4}$ $\frac{5}{8}$ $\frac{1}{5}$ $\frac{6}{10}$ _____ _____ _____ _____

27. $.4$ $\frac{2}{10}$ $.04$ $\frac{1}{4}$ _____ _____ _____ _____

Read the sentences, then write the fractions as percents.

28. Marge bought the shoes at a half-price sale.   _____

29. Bill's rent is $\frac{1}{5}$ of his monthly salary.   _____

30. His salary raise was $\frac{1}{4}$ of his base salary.   _____

Meaning of Percent

A **percent** is a ratio that compares a number to 100. Percent also means *hundredths,* or *per hundred.* Percents are a common way of talking about money, interest on loans, and payroll deductions. Percents are also used to describe the relationship between parts and whole units. For example, you can use percents to specify the amount of ingredients in a package or participants in an event.

The symbol for percent is **%.** 100% means one whole unit or the entire amount. 0% means none.

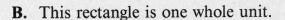

Examples

A. One dollar is like one whole unit.

B. This rectangle is one whole unit.

Divide one dollar into 100 parts.

Divide one whole unit into 100 parts.

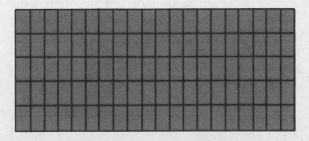

One dollar equals 100 cents, or 100% of the whole unit.
$1.00 = 100¢

One whole unit equals 100 percent.
1 whole unit = 100%

29

C. The whole unit is a T-shirt. The label says 100% cotton. This means that only cotton was used to make the T-shirt.

D. The whole unit is the corn oil. The label says 100% corn oil. This means that the only ingredient in the bottle is corn oil.

E. The whole unit is the luncheon meat. 0% fat means there is no fat in it.

Practice

Refer to the pictures and answer the questions.

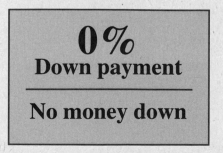

1. (1) What is the whole unit?

_____

   (2) What does 100% natural mean?

_____

2. (1) What is the whole unit?

_____

   (2) What does 0% down payment mean?

_____

3. (1) What is the whole unit?

_____

(2) What does 100% of the minimum daily requirement mean?

_____

4.

What percent of the oil has cholesterol?

_____

5.

Vote of building committee		
	Yes	No
Anderson	×	
Klein	×	
O'Brien	×	
Pereg	×	

What percent of the committee voted YES?

_____

6.

July					
1̶	2̶	3̶	4̶	5̶	6̶
7̶	8̶	9̶	1̶0̶	1̶1̶	1̶2̶	1̶3̶
1̶4̶	1̶5̶	1̶6̶	1̶7̶	1̶8̶	1̶9̶	2̶0̶
2̶1̶	2̶2̶	2̶3̶	2̶4̶	2̶5̶	2̶6̶	2̶7̶
2̶8̶	2̶9̶	3̶0̶	3̶1̶		

What percent of the month is marked off?

_____

7.

The sign on the theater's ticket window read "Sold Out." What percent of the tickets were sold?

_____

Solve the following problems.

8. The label on the muffin mix says that it is salt free. What percent of salt is in the muffin mix?

_____

9. The label on the paper plates states that they are 100% recyclable. What percent of the paper plates can be recycled?

_____

10. The entire class passed the GED exam. What percent of the class passed the test?

_____

11. Everyone preferred pizza for dinner. What percent of the group preferred pizza?

_____

12. The sweater's label says 100% lamb's wool. What percent of the sweater is made from lamb's wool?

_____

13. The cold drink needed 100% cold water. What percent of cold water is in the drink?

_____

14. The white paint mix has 100% white paint. What percent of white paint is in the paint mix?

_____

15. The entire family went to the movies. What percent of the family went to the movies?

_____

Using Nutritional Information From Labels

Phyllis wanted to be certain the foods she ate provided at least 100% of the recommended daily allowance (RDA) of calcium. She checked the nutritional information on the labels of the food. The labels give the percent of the RDA in one serving of the food.

Here is the chart Phyllis made.

One serving	% RDA calcium
Cold cereal with milk	20
Yogurt	30
Sardines	30
Ice cream	30
Swiss cheese	25
Pudding	15
Cream of tomato soup	15

Here is what Phyllis ate in one day.

These foods	% RDA calcium
Cereal with milk	20
Yogurt	30
Sardines	30
Ice cream	30
Adding all the percents, Phyllis came up with 110%.	110%

What would you eat in order to get at least **100% of the RDA of iron?** Use the information below. Make a list and be sure the percents add up to **at least** 100%.

One serving	% RDA iron	These foods	% RDA iron
Cold cereal with milk	15	_____	_____
Hot farina cereal	45	_____	_____
Natural wheat bread	20	_____	_____
Beans	25	_____	_____
Spaghetti	25	_____	_____
Rice pudding	5	_____	_____
		Total	_____

33

Meaning of Percents Between 0 and 100

Percent means that some amount or unit has been divided into **100 parts.** Percent refers to how many parts of 100 you are talking about.

One hundred cents equals one whole dollar.

$$100¢ = \$1.00$$

One hundred percent equals one whole.

$$100\% = \text{one whole unit}$$

> **MATH HINT**
>
> **P**ercents are expressed in parts per 100. 1% is $\frac{1}{100}$, or .01, of a unit. 1%, .01, and $\frac{1}{100}$ are all equal.

This shaded penny is one cent out of 100 cents in a whole dollar.

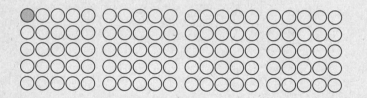

It can be written as **$.01** or **1¢.**

One cent is also $\frac{1}{100}$ of a dollar.

One cent is **1%** of a dollar.

This shaded square is one out of 100 squares in this rectangle.

It can be written as **.01** of the rectangle.

One square is also $\frac{1}{100}$ of this rectangle.

One square is **1%** of the rectangle.

A. 98 cents is 98% of a dollar.

Below, one dollar is shown as 100 pennies.

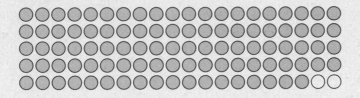

How many pennies are shaded?
98 pennies

What percent of the dollar is shaded?
98% of the dollar

B. 98 parts out of 100 can be written as 98%.

Below, the rectangle has been divided into 100 squares.

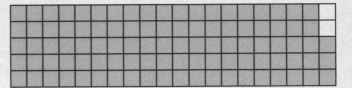

How many parts of the rectangle are shaded?
98 squares

What percent of the rectangle is shaded?
98% of the rectangle

Each figure below has been divided into **100 parts.** Study the shaded and unshaded parts to determine the answers to the questions.

This circle is divided into 100 parts.

75 parts are shaded.

25 parts are not shaded.

1. What percent of the circle is shaded?

_____

2. What percent of the circle is not shaded?

_____

3. What percent of the rectangle is shaded?

   _____

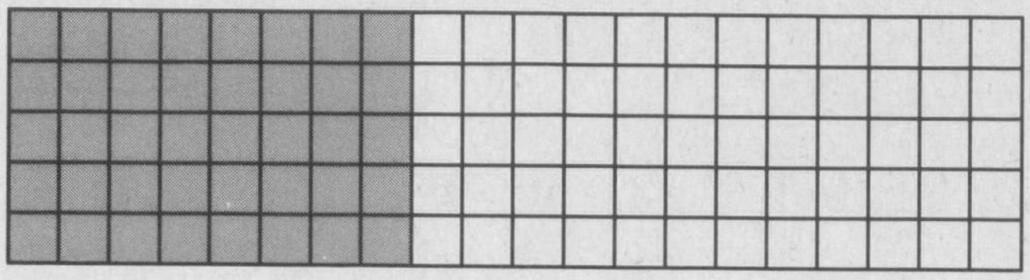

4. What percent of the rectangle is not shaded?

   _____

5. What is the total percent shaded and not shaded in the rectangle?

   _____

6. What percent of the large triangle is shaded?

   _____

The large triangle is divided into 100 parts.

7. What percent of the large triangle is not shaded?

   _____

8. What is the total percent shaded and not shaded in the large triangle?

   _____

9. What percent of the box is shaded?

   _____

The box is divided into 100 equal parts.

10. What percent of the box is not shaded?

    _____

11. What is the total percent shaded and not shaded in the box?

    _____

36

12. What percent of the circle is shaded?

_____

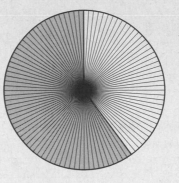

The circle is divided into 100 equal parts.

13. What percent of the circle is not shaded?

_____

14. What is the total percent shaded and not shaded in the circle?

_____

15. Explain in your own words why the total percent shaded and not shaded has the same answer for every question.

_____

_____

_____

_____

Meaning of Less Than 1%

Less than 1% means less than $\frac{1}{100}$ of a whole unit.

Half a cent is like half a percent.

.5 cent, or $\frac{1}{2}$ cent

The box is divided into 100 parts.

$99\frac{1}{2}$ of the parts are not shaded.

$99\frac{1}{2}$% of the box is not shaded.

(Note: 99.5% is the same as $99\frac{1}{2}$%.)

$\frac{1}{2}$ of one of the parts is shaded.

$\frac{1}{2}$%, or .5%, of the box is shaded.

─────────── **Example** ───────────

Using the figure below, read the questions and answers.

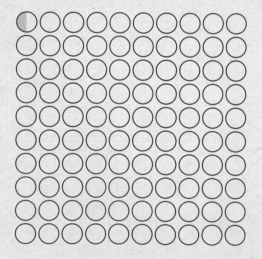

$\frac{1}{2}$% of a dollar is $\frac{1}{2}$ penny.

How many pennies are shaded?
$\frac{1}{2}$ penny

How many pennies are not shaded?
$99\frac{1}{2}$ pennies

What percent of the dollar is shaded?
$\frac{1}{2}$%

What percent of the dollar is not shaded?
$99\frac{1}{2}$%

Answer the questions.

1. Box A is divided into 100 parts.
   $\frac{3}{4}$ of one of the parts is shaded.
   $99\frac{1}{4}$ of the parts are not shaded.

   Box A

   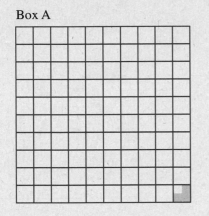

   (1) What percent of the box is not shaded?

   _____

   (2) What percent of the box is shaded?

   _____

   (3) What percent does the whole box represent?

   _____

2. Box B is divided into 100 parts.
   .75 of one of the parts is shaded.
   99.25 of the parts are not shaded.

   Box B

   (1) What percent of the box is not shaded?

   _____

   (2) What percent of the box is shaded?

   _____

   (3) Is there any difference between Box A and Box B?

   _____

   (4) Explain in your own words why or why not.

   _____

   _____

   (5) Which is larger, Box A or Box B?

   _____

Solve the following problems.

3. For every dollar paid in taxes, $\frac{1}{2}$ penny is spent on the park district. Write the amount spent as a decimal percent.

   _____

4. For every dollar paid on gasoline tax, $\frac{1}{5}$ cent goes to cleaning the highways. Write the amount spent to clean highways as a fraction percent.

   _____

5. For every dollar paid in taxes, 8 cents is spent on roads. Write the amount spent on roads as a percent.

   _____

6. If you know that .4% of a box is shaded, how can you calculate what percent is not shaded?

   _____

7. For every dollar spent on taxes, 9 cents goes to schools. Write the amount that goes to schools as a percent.

   _____

8. Define less than 1% in your own words.

   _____

   _____

   _____

Meaning of More Than 100%

More than 100% means more than one whole unit. For example, two dollars are like two whole units.

Each dollar equals 100 cents.
Together both dollars equal 200 cents.
$2.00 = 200¢.
$2.00 is 200% of one dollar.

Each box has 100 squares.
Together both boxes have 200 squares.
2 whole boxes = 200%.
2 whole boxes is 200% of one box.

Examples

A. What is 300% of one dollar?

300% of one dollar is $3.00.

B. What is 350% of this box?

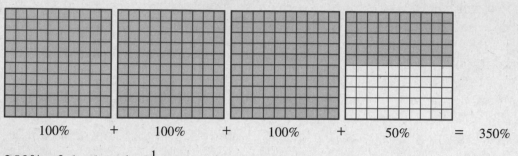

100%   +   100%   +   100%   +   50%   =   350%

350% of the box is $3\frac{1}{2}$ boxes.

Practice

Represent the following percentages by shading the figures below.

1. **Shade in 250% of one box.**

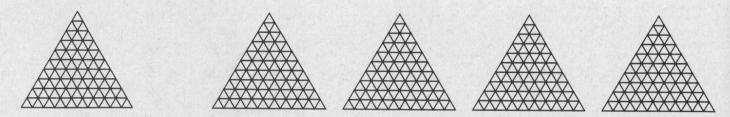

2. **Shade in 350% of one triangle.**

42

Understanding Percent Problems

Percent problems are used to relate different numbers to each other in terms of 100.

100% of a number is **one** times the number.

200% is the same as **two** times the number. The words **twice** the amount, **two times** the amount, **double** the amount, and **200%** of the amount all mean the same.

300% is **three** times the amount. It means the same as **triple** the amount.

400% is **four** times the amount. It means the same as **quadruple** the amount.

Example

The cost of living today is **four times** what it was in 1970. Write the percent equivalent of the bold-faced phrase.

Four times an amount is the same as 400% of an amount. Therefore, the cost of living today is **400%** of what it was in 1970.

Practice

Write the percent equivalent to the following phrases on the blanks provided.

1. Double your money back _____

2. Triple value on your coupons _____

3. Three times the value _____

4. Four times the fun _____

5. Twice the return on your investment _____

6. Quadruple your money _____

7. Five times as likely to flood _____

8. Double the rate per capita _____

9. Three times more likely to try _____

10. Three times the national average _____

Estimating Percents

To estimate means to find an answer that is close to the exact answer. Visualizing a percent problem often helps to understand it. By using the figures below, you can estimate the value of a percent.

0%   100%

Step 1   Draw a line down the middle of the figure. That is 50%.

0%   50%   100%

Step 2   Double the figure. That is 200%.

0%   50%   100%   200%

Once you have found 50%, 100%, and 200%, then it is easier to estimate a given percent.

──────────────── **Example** ────────────────

Shade in 145% using the figure below.

Step 1   Draw a line down the middle. Mark it 50%.

0%   50%   100%

Step 2   Double the figure. Mark it 200%.

0%   50%   100%   150%   200%

Shade in 145%. It will be just a little less than 150%.

Draw a line down the middle of each figure, then double it. Mark off
50%, 150%, and 200%; then shade in the percentage given.

1. 45%

2. 75%

3. 195%

4. 5%

5. 80%

6. 132%

7. 100%

8. 175%

9. 20%

10. 60%

Percents and the Decimal Number Line

Percents can be explained using the number line.

Look at the number line in the example.

The top part of the number line is marked off in decimals from zero to one.

The bottom of the number line is marked off into percentages.

Example

0	.25	.50	.75	1.00
0	25%	50%	75%	100%

Look at the number line above.
The top part of the number line shows .50.
Find the percent written directly below.

The bottom part of the number line shows 50%.
50% means 50 parts of 100.
.50 simplifies to .5.
.5 equals 50%.

This number line is divided by decimals from zero to one.

0	.1	.2	.25	.3	.4	.5	.6	.7	.75	.8	.9	1.0
0	10%	20%	25%	30%	40%	50%	60%	70%	75%	80%	90%	100%

In problems 1–7, use the number line above to show the relationship between the following percents and decimals. Write the decimal equivalent of each percentage on the line provided.

1. 25% _____

2. 75% _____

3. 40% _____

4. 80% _____

5. 30% _____

6. 90% _____

7. 20% _____

In problems 8–14, use the number line above to show the relationship between the following decimals and percents. Write the percent equivalent of each decimal on the line provided.

8. 1 _____

9. .25 _____

10. .2 _____

11. .7 _____

12. .9 _____

13. .6 _____

14. .1 _____

15. Describe in your own words the relationship between decimals and percents.

_____

_____

Percents and the Fraction Number Line

Percents can also be explained using the number line.

Look at the number line in the example.
The top part of the number line is marked off in fractions
from zero to one.

The bottom of the number line is marked off into percentages.

Example

This number line is divided by fractions from zero to one.

$$0 \quad \frac{1}{4} \quad \frac{1}{2} \quad \frac{3}{4} \quad 1$$
$$0 \quad \frac{25}{100} \quad \frac{50}{100} \quad \frac{75}{100} \quad \frac{100}{100}$$
$$0 \quad 25\% \quad 50\% \quad 75\% \quad 100\%$$

Look at the number line above.
The top part of the number line shows $\frac{50}{100}$.
Find the percent written directly below.

The bottom part of the number line shows 50%.
50% means 50 parts of 100.
$\frac{50}{100}$ **reduces to** $\frac{1}{2}$.
$\frac{1}{2}$ **equals** 50%.

| 0 | $\frac{1}{10}$ $\frac{10}{100}$ | $\frac{1}{5}$ $\frac{20}{100}$ | $\frac{1}{4}$ $\frac{25}{100}$ | $\frac{3}{10}$ $\frac{30}{100}$ | $\frac{2}{5}$ $\frac{40}{100}$ | $\frac{1}{2}$ $\frac{50}{100}$ | $\frac{3}{5}$ $\frac{60}{100}$ | $\frac{7}{10}$ $\frac{70}{100}$ | $\frac{3}{4}$ $\frac{75}{100}$ | $\frac{4}{5}$ $\frac{80}{100}$ | $\frac{9}{10}$ $\frac{90}{100}$ | 1 $\frac{100}{100}$ |

| 0 | 10% | 20% 25% 30% | 40% | 50% | 60% | 70% 75% 80% | 90% | 100% |

In problems 1–7, use the number line above to show the relationship between the following percents and fractions. Write the fraction equivalent of each percentage on the lines provided.

1. 25% _____

2. 75% _____

3. 40% _____

4. 80% _____

5. 30% _____

6. 90% _____

7. 20% _____

In problems 8–14, use the number line above to show the relationship between the following fractions and percents. Write the percent equivalent of each fraction on the lines provided.

8. $\frac{1}{2}$ _____

9. $\frac{1}{4}$ _____

10. $\frac{1}{5}$ _____

11. $\frac{4}{5}$ _____

12. $\frac{2}{5}$ _____

13. $\frac{7}{10}$ _____

14. $\frac{3}{10}$ _____

15. Describe in your own words the relationship between fractions and percents.

_____

_____

Buying With Discounts, Rebates, or Coupons

A **discount** is an amount of money subtracted from the original price.

A **rebate** is a refund of money given after something has been purchased.

A **coupon** is a written statement giving a discount on the price of something.

Tito and Anya visited two stores. Both Ranji's Furniture Store and Martinez Furniture Mart sell the same brands of furniture at the same price. However each store offered a furniture sale.

Circle the number of the store offering the better discount. (**Hint:** Change the fractional discounts to percentages.)

1. Two love seats
   (1) 39% discount at Ranji's
   (2) $\frac{1}{2}$ off the price at Martinez's

2. Dining room set
   (1) $33\frac{1}{3}\%$ discount at Ranji's
   (2) $\frac{1}{4}$ off the price at Martinez's

3. Lamp
   (1) 10% discount at Ranji's
   (2) $\frac{1}{4}$ off the price at Martinez's

Mariko and Yukio want to purchase car accessories. Kosinki's Auto Parts gives rebates. Goldberg's Auto Supply publishes coupons.

If the prices and the brands are identical, circle the number of the store offering the better buy.

4. Spark plugs
   (1) 10% rebate at Kosinki's
   (2) $\frac{1}{3}$ off coupon at Goldberg's

5. Antifreeze
   (1) 60% rebate at Kosinki's
   (2) $\frac{1}{2}$ off coupon at Goldberg's

6. Windshield wipers
   (1) 30% rebate at Kosinki's
   (2) $\frac{1}{4}$ off coupon at Goldberg's

Writing Equal Fractions, Decimals, and Percents

Percents are often used to discuss money, taxes, discounts, or interest on loans. However, you cannot multiply or divide by percents. Before you can solve a percent problem, you must change the percent to a fraction or a decimal. Fractions, decimals, and percents are different ways to write the **same value.**

Examples

A. Here are different ways you can write **one quarter.**

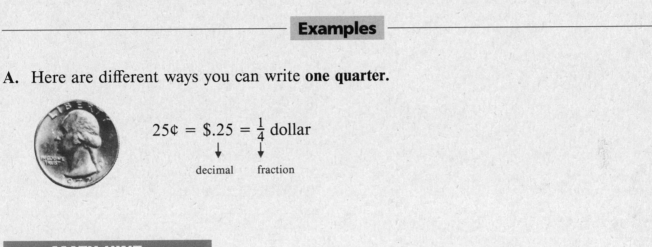

$$25¢ = \$.25 = \tfrac{1}{4} \text{ dollar}$$

decimal   fraction

MATH HINT

To write $.25 in fraction form, do this:

$\$.25 = \dfrac{25}{100}$ or $\tfrac{1}{4}$ dollar.

B. Here are different ways you can write 25%.

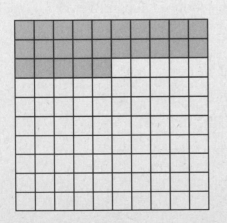

$$\frac{25}{100} = \tfrac{1}{4} = .25 = 25\%$$

fraction   decimal   percent

Write the answers.

1. Write half a dollar as a fraction.

   _____

2. Write five tenths as a decimal.

   _____

3. Write half a dollar with a ¢ sign.

   _____

4. Write a dime as a decimal.

   _____

5. Write half a dollar with a $ sign.

   _____

6. Write a dime with a ¢ sign.

   _____

7. Write a dime as a fraction of a dollar.

   _____

8. Write a dime with a $ sign.

   _____

9. Write $\frac{1}{2}$ as a percent.

   _____

10. Write $\frac{3}{4}$ as a percent.

    _____

11. One-half of the 100 people working in a factory have a savings plan. What percent of workers have a savings plan?

    _____

12. Three-fourths of the cars that came into a gas station used unleaded gasoline. What percent of cars used unleaded gasoline?

    _____

Use the figures to answer the following questions.

13. Shade $\frac{3}{4}$ of the box.

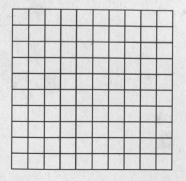

14. Shade .75 of the box.

15. Shade 75% of the box.

16. Compare $\frac{3}{4}$, .75, and 75%. Describe the relationship in your own words.

_____

_____

_____

17. Shade $\frac{1}{5}$ of the box.

18. Shade .2 of the box.

19. Shade 20% of the box.

20. Compare $\frac{1}{5}$, .2, and 20%. Describe the relationship in your own words.

_____

_____

_____

Changing Decimals to Percents

Decimals and percents are two ways to express the same amount.
To change decimals to percents, follow these steps:

Step 1 Multiply the decimal by 100.

Step 2 Add the % sign.

> **MATH HINT**
>
> To multiply by 100, move the decimal point two places to the right.
>
> $.75 = .75 = 75\%$

Examples

A. Change .5 to a percent.

$$\begin{array}{r} .5 \\ \times 100 \\ \hline 50.0 \end{array}$$

Step 1 Multiply the decimal times 100.

Step 2 Add the percent sign.

50%

B. Change 4 to a percent.

$$\begin{array}{r} 4.0 \\ \times 100 \\ \hline 400.0 \end{array}$$

Step 1 Multiply the decimal times 100.

Step 2 Add the percent sign.

400%

C. Change .005 to a percent.

$$\begin{array}{r} .005 \\ \times 100 \\ \hline .500 \end{array}$$

Step 1 Multiply the decimal times 100.

Step 2 Add the percent sign.

.5%

Write the following decimals as percents.

1. .35 _____   2. .01 _____

3. 9. _____   4. .002 _____

5. .9 _____   6. .1 _____

Problem Solving

Read the sentences, then write the decimals as percents.

7. Gasoline tax increased by .005.

_____

8. This drink is only .1 fruit juice.

_____

9. The 1.75 increase in the cost of health insurance was paid by the company.

_____

10. .85 of the graduating class decided to search for a job.

_____

11. The soft drink contained .2 sodium.

_____

12. The shirt was made of .60 cotton.

_____

13. The cookies had .30 butter in them.

_____

14. The jelly had .75 strawberries in it.

_____

Changing Percents to Decimals

Decimals and percents are two ways to express the same amount. To change percents to decimals, follow these steps:

Step 1   Remove the % sign.

Step 2   Multiply by .01.

> **MATH HINT**
>
> To multiply by .01, move the decimal point two places to the left.
>   75% = 75.0% = .75
> The result is the same as dividing by 100.

Examples

A. Change 5% to a decimal.

$$\begin{array}{r} 5 \\ \times\,.01 \\ \hline .05 \end{array}$$

.05

Step 1   Remove the % sign.
Step 2   Multiply by .01.

B. Change 400% to a decimal.

$$\begin{array}{r} 400 \\ \times\quad.01 \\ \hline 4.00 \end{array}$$

4

Step 1   Remove the % sign.
Step 2   Multiply by .01.

C. Change .5% to a decimal.

$$\begin{array}{r} .5 \\ \times\quad.01 \\ \hline .005 \end{array}$$

.005

Step 1   Remove the % sign.
Step 2   Multiply by .01.

Write the following percents as decimals.

1. 35% _____

2. 198% _____

3. 2% _____

4. 59% _____

5. 999% _____

6. .7% _____

Read the sentences below. Then write the percents as decimals.

7. Five percent of Leo's salary is withheld for dental insurance. _____

8. The price of lettuce has increased 140% over the past 10 years. _____

9. Ninety percent of the people attending the conference were self-employed. _____

10. Seventy-five percent of the paper used at the company was recycled. _____

11. Margie saves 19% of her salary. _____

12. Thirty-two percent of the graduating class had jobs one month after graduation. _____

13. There was a 50% rebate coupon available for the new shampoo. _____

14. Sixty percent of the visitors to the park were children under ten years of age. _____

15. Eighty-three percent of the employees take their vaction in June. _____

Changing Percents to Fractions and Mixed Numbers

Fractions and percents are two ways to express the same amount.
To change fractions to percents, follow these steps:

Step 1 Remove the percent (%) sign. Change to an improper fraction if necessary.

Step 2 Multiply by $\frac{1}{100}$.

Step 3 Cancel and reduce if necessary.

Examples

A. Change 8% to a fraction.

$$8 = \frac{8}{1}$$

Step 1 Remove the % sign. Change to an improper fraction.

$$\frac{8}{1} \times \frac{1}{100}$$

Step 2 Multiply the fraction times $\frac{1}{100}$.

$$\frac{\overset{2}{\cancel{8}}}{1} \times \frac{1}{\underset{25}{\cancel{100}}} = \frac{2}{25}$$

Step 3 Cancel and reduce.

B. Change 975% to a fraction.

$$975 = \frac{975}{1}$$

Step 1 Remove the % sign. Change to an improper fraction.

$$\frac{975}{1} \times \frac{1}{100}$$

Step 2 Multiply the fraction times $\frac{1}{100}$.

$$\frac{\overset{39}{\cancel{975}}}{1} \times \frac{1}{\underset{4}{\cancel{100}}} = \frac{39}{4} = 9\frac{3}{4}$$

Step 3 Cancel and reduce.

C. Change $\frac{1}{2}$% to a fraction.

$$\frac{1}{2}$$

Step 1 Remove the % sign.

$$\frac{1}{2} \times \frac{1}{100} = \frac{1}{200}$$

Step 2 Multiply the fraction times $\frac{1}{100}$.

D. Change $33\frac{1}{3}$% to a fraction.

$$33\frac{1}{3} = \frac{100}{3}$$

Step 1 Remove the % sign. Change the mixed number to an improper fraction.

$$\frac{100}{3} \times \frac{1}{100}$$

Step 2 Multiply by $\frac{1}{100}$.

$$\frac{\overset{1}{\cancel{100}}}{3} \times \frac{1}{\underset{1}{\cancel{100}}} = \frac{1}{3}$$

Step 3 Cancel and reduce.

Write the following percents as fractions. Reduce to lowest terms.

1. 90%   _____
2. 13%   _____
3. 190%   _____

4. 22%   _____
5. 60%   _____
6. 85%   _____

7. 150%   _____
8. 50%   _____
9. 25%   _____

10. 100%   _____
11. $66\frac{2}{3}$%   _____
12. $37\frac{1}{2}$%   _____

13. $16\frac{2}{3}$%   _____
14. $87\frac{1}{2}$%   _____
15. $83\frac{1}{3}$%   _____

16. $16\frac{1}{2}$%   _____
17. $21\frac{1}{4}$%   _____
18. $8\frac{1}{3}$%   _____

Problem Solving

Read the sentences, then write the percents as fractions. Reduce to lowest terms.

19. Tables are on sale for 25% off.   _____

20. Ground beef is marked 80% lean.   _____

21. Seventy-five percent of the clients were return customers.   _____

22. Twenty percent of the phone calls were sales calls.   _____

23. There is a 300% markup on the cost of electronics equipment.   _____

24. Sixty percent of the class passed the final.   _____

25. The library showed a loss of 20% of its children's books this year.   _____

Changing Fractions and Mixed Numbers to Percents

Fractions and percents are two ways to express the same amount.
To change fractions to percents, follow these steps:

Step 1   Multiply the fraction by $\frac{100}{1}$.
Step 2   Cancel and reduce.
Step 3   Write any remainders as fractions.
Step 4   Add the % sign.

Examples

A.   Change $\frac{1}{2}$ to a percent.

$(\frac{1}{2} \times \frac{100}{1})$ **Step 1**   Multiply the fraction by $\frac{100}{1}$.

$(\frac{1}{\cancel{2}} \times \frac{\overset{50}{\cancel{100}}}{1}) = \frac{50}{1}$ **Steps 2 and 3**   Cancel and reduce. Make any remainders into fractions.

50%   **Step 4**   Add the % sign.

B.   Change $2\frac{1}{4}$ to a percent.

$2\frac{1}{4} \times \frac{100}{1}$ **Step 1**   Multiply by $\frac{100}{1}$.

$\frac{9}{\cancel{4}} \times \frac{\overset{25}{\cancel{100}}}{1} = \frac{225}{1}$ **Steps 2 and 3**   Cancel and reduce. Make any remainders into fractions.

225%   **Step 4**   Add the % sign.

> **MATH HINT**
>
> **R**emember, a mixed fraction must be converted into an improper fraction when multiplying.
> $$2\frac{1}{5} = \frac{11}{5}$$

C.   Change $\frac{1}{500}$ to a percent.

$\frac{1}{500} \times \frac{100}{1}$ **Step 1**   Multiply by $\frac{100}{1}$.

$\frac{1}{\underset{5}{\cancel{500}}} \times \frac{\overset{1}{\cancel{100}}}{1} = \frac{1}{5}$ **Steps 2 and 3**   Cancel and reduce. Make any remainders into fractions.

$\frac{1}{5}$%   **Step 4**   Add the % sign.

D. Change $\frac{2}{3}$ to a percent.

$\frac{2}{3} \times \frac{100}{1}$ **Step 1**   Multiply by $\frac{100}{1}$.

$\frac{200}{3} = 66\frac{2}{3}$ **Steps 2 and 3**   Reduce.

$66\frac{2}{3}\%$ Make any remainders into fractions.

                  Step 4   Add the % sign.

Practice

Write the following fractions as percents.

1. $\frac{2}{5}$ _____   **2.** $\frac{1}{4}$ _____   **3.** $\frac{1}{25}$ _____

4. $\frac{29}{10}$ _____   **5.** $\frac{3}{20}$ _____   **6.** $\frac{1}{3}$ _____

7. $\frac{5}{6}$ _____   **8.** $\frac{3}{8}$ _____   **9.** $\frac{1}{6}$ _____

10. $\frac{5}{8}$ _____   **11.** $\frac{7}{8}$ _____   **12.** $\frac{1}{12}$ _____

13. $\frac{1}{7}$ _____   **14.** $\frac{5}{7}$ _____   **15.** $\frac{6}{7}$ _____

Problem Solving

Read the sentences; then write the fractions as percents.

16. The cost of health insurance is increased by $\frac{1}{5}$. _____

17. The crime rate decreased by $\frac{3}{4}$. _____

18. The cost of living has increased $3\frac{1}{2}$ times in the past 20 years. _____

19. The material in the shirt was $\frac{4}{5}$ silk. _____

20. The interest rate is $\frac{4}{25}$ of the amount owed. _____

Practice Writing Equal Fractions, Decimals, and Percents

The same amount can be expressed as a fraction, a decimal, or a percent.

Recall the steps for changing percents to fractions:

Step 1 Remove the % sign.
Step 2 Multiply by $\frac{1}{100}$.
Step 3 Cancel and reduce if necessary.

Recall the steps for changing percents to decimals:

Step 1 Remove the % sign.
Step 2 Multiply by .01.

Recall the steps for changing fractions to percents:

Step 1 Multiply by $\frac{100}{1}$.
Step 2 Cancel and reduce.
Step 3 Make any remainders into fractions.
Step 4 Add the % sign.

Recall the steps for changing decimals to percents:

Step 1 Multiply the decimal by 100.
Step 2 Add the % sign.

Examples

A. Express 50% as a fraction.

$$50\% = \frac{\overset{1}{\cancel{50}}}{1} \times \frac{1}{\underset{2}{\cancel{100}}} = \frac{1}{2}$$

B. Express 50% as a decimal.

$$\begin{array}{r} 50 \\ \times\ .01 \\ \hline .50 \end{array}$$

50%, $\frac{1}{2}$, and .50 are different ways to express the same amount.

Fill in the chart by changing the given percent, fraction, or decimal to the other forms.

Percent	Fraction	Decimal
1. $33\frac{1}{3}\%$	_____	_____
2. _____	$\frac{1}{10}$	_____
3. $16\frac{2}{3}\%$	_____	_____
4. _____	_____	1.35
5. _____	$\frac{5}{8}$	_____
6. _____	_____	.90
7. _____	$\frac{1}{500}$	_____
8. $1\frac{1}{2}\%$	_____	_____
9. _____	$\frac{99}{100}$	_____
10. 2,557%	_____	_____
11. _____	$\frac{1}{5}$	_____
12. 75%	_____	_____
13. _____	_____	$.66\frac{2}{3}$
14. _____	$\frac{1}{8}$	$.12\frac{1}{2}$
15. 60%	_____	_____

Reading Ads

The Costellos are reading the Sunday paper's advertisements. They are trying to find the best deals on electronics equipment. Some of the ads use fractions and some use percents to advertise their discounts.

Read the following ads. Match the ads using the percent with the price tag using the equivalent fraction discount.

1.

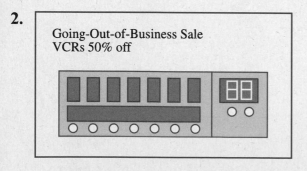

Prices slashed!!!
washer and dryer combination
33 1/3%

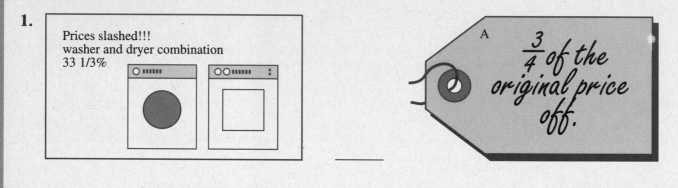

A

$\frac{3}{4}$ of the original price off.

2.

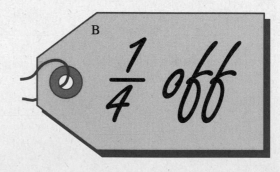

Going-Out-of-Business Sale
VCRs 50% off

B

$\frac{1}{4}$ off

3.

Moving! Everything must go!
75% reduction on all TVs

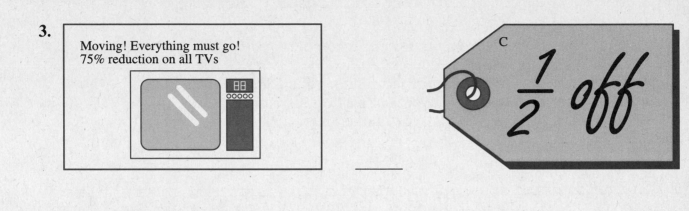

C

$\frac{1}{2}$ off

_____

4.

President's Day Sale!!
25% off the cost of every camera in the store

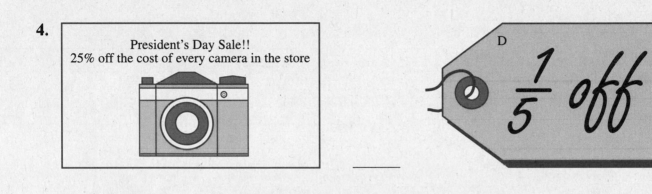

D

$\frac{1}{5}$ off

_____

5.

Computers for the home.
This weekend only
20% off

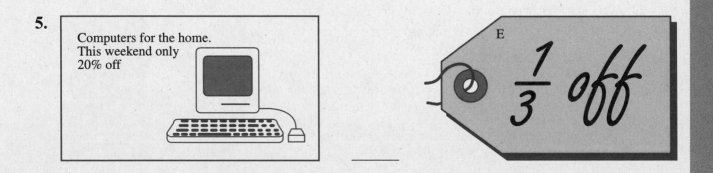

E

$\frac{1}{3}$ off

_____

Using Number Lines to Compare Percents to Decimals and Fractions

Decimals and percents can be compared using a number line.

Look at number line **A**. The **top part** of this number line is marked off in decimals from zero to one. The **bottom part** of this number line is marked off in percents. Each of the marked increments represents 10%.

A.

0	.1	.2	.3	.4	.5	.6	.7	.8	.9	1
0	10%	20%	30%	40%	50%	60%	70%	80%	90%	100%

Fractions and percents can be compared using a number line.

Look at number line **B**. The **top part** of this number line is marked off in fractions from zero to one. The **bottom part** is marked off in percents. Each of the marked increments represents 10%.

B.

0	$\frac{1}{10}$	$\frac{1}{5}$	$\frac{3}{10}$	$\frac{2}{5}$	$\frac{1}{2}$	$\frac{3}{5}$	$\frac{7}{10}$	$\frac{4}{5}$	$\frac{9}{10}$	1
0	$\frac{10}{100}$	$\frac{20}{100}$	$\frac{30}{100}$	$\frac{40}{100}$	$\frac{50}{100}$	$\frac{60}{100}$	$\frac{70}{100}$	$\frac{80}{100}$	$\frac{90}{100}$	$\frac{100}{100}$
0	10%	20%	30%	40%	50%	60%	70%	80%	90%	100%

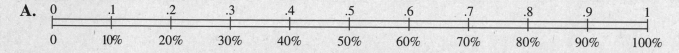

Examples

A. Which is larger, .5 or 30%?

Step 1 Using number line A, locate .5 along the top of the number line.

Step 2 Find 30% along the bottom of number line A.

Step 3 The value farther to the right is the larger.

.5 is to the right of 30%.

Therefore, .5 > 30%.

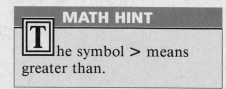

MATH HINT

The symbol > means greater than.

B. Which is larger, $\frac{1}{5}$ or 30%?

Step 1 Using number line B, find $\frac{1}{5}$ along the top of the number line.

Step 2 Find 30% along the bottom of number line B.

Step 3 The value farther to the right is the larger.

30% is to the right of $\frac{1}{5}$.
Therefore, 30% > $\frac{1}{5}$.

Practice

Using the number lines below, compare decimals, fractions, and percents.

0	$\frac{1}{10}$	$\frac{1}{5}$	$\frac{1}{4}$	$\frac{3}{10}$	$\frac{2}{5}$	$\frac{1}{2}$	$\frac{3}{5}$	$\frac{7}{10}$	$\frac{3}{4}$	$\frac{4}{5}$	$\frac{9}{10}$	1
0	10%	20%	**25%**	30%	40%	**50%**	60%	70%	**75%**	80%	90%	100%

0	.1	.2	.3	.4	.5	.6	.7	.8	.9	1
0	10%	20%	30%	40%	50%	60%	70%	80%	90%	100%

Which is larger?

1. 75% or .8 _____

2. $\frac{1}{10}$ or 15% _____

3. .22 or 1% _____

4. 66% or $\frac{3}{4}$ _____

5. 98% or 1 _____

6. 90% or 1 _____

7. 18% or .6 _____

8. 25% or $\frac{1}{2}$ _____

9. 33% or .3 _____

10. $\frac{4}{5}$ or 85% _____

11. 100% or .9 _____

12. $\frac{9}{10}$ or 99% _____

Which is smaller?

13. .1 or 5% _____

14. $\frac{2}{5}$ or 45% _____

15. 55% or .7 _____

16. $\frac{3}{10}$ or 20% _____

17. 14% or .2 _____

18. $\frac{1}{2}$ or 49% _____

19. 99% or .3 _____

20. 19% or $\frac{1}{10}$ _____

Comparing Fractions, Decimals, and Percents

Fractions, decimals, and percents are different ways to write numbers. Unless you memorize all the equivalent fractions, decimals, and percents, the best way to compare the value of those numbers is to put them all into the same form.

Examples

A. Compare $\frac{3}{4}$, .88, and 80%.
Write these numbers in order from largest to smallest.

Decimal Method
Change all the values to decimals.
Expand each number to the same decimal place value.

$$\frac{3}{4} = 4\overline{)3.00} = .75$$

$$.88 = .88$$

$$80\% = 80 \times .01 = .80$$

Compare the place value.

Largest .88

Middle .80

Smallest .75

Put in order from largest to the smallest.

$$.88 \quad > \quad 80\% \quad > \quad \frac{3}{4}$$

B. Compare $\frac{3}{10}$, 35%, and .38.
Write these numbers in order from the smallest to the largest.

Fraction Method
Change all the values to fractions.
Find the common denominator for each fraction.

$$\frac{3}{10} \quad = \quad \frac{3}{10} \times \frac{10}{10} \quad = \quad \frac{30}{100}$$

$$35\% \quad = \quad \frac{35}{1} \times \frac{1}{100} \quad = \quad \frac{35}{100}$$

$$.38 \quad = \quad \frac{38}{100} \quad = \quad \frac{38}{100}$$

Compare numerators.

Smallest $\frac{30}{100}$

Middle $\frac{35}{100}$

Largest $\frac{38}{100}$

Put in order from smallest to the largest.

$$\frac{3}{10} \quad < \quad 35\% \quad < \quad .38$$

MATH HINT

Remember, fractions mean to divide the numerator by the denominator. Mixed numbers must be changed to improper fractions first.

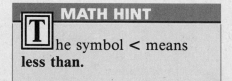

MATH HINT

The symbol < means **less than.**

Change each group of numbers to decimals. Then, write the numbers
in order from the largest to the smallest.

1. $\frac{1}{10}$ .09 2%

   (1) $\frac{1}{10}$ = _____

   (2) .09 = _____

   (3) 2% = _____

   (4) ____ > ____ > ____

2. 100% $1\frac{1}{2}$ 1.45

   (1) 100% = _____

   (2) $1\frac{1}{2}$ = _____

   (3) 1.45 = _____

   (4) ____ > ____ > ____

3. .33 3.5% $\frac{3}{10}$

   (1) .33 = _____

   (2) 3.5% = _____

   (3) $\frac{3}{10}$ = _____

   (4) ____ > ____ > ____

Change each group of numbers to fractions. Then, write the numbers
from the smallest to the largest.

4. 55% .6 $\frac{7}{10}$

   (1) 55% _____

   (2) .6 _____

   (3) $\frac{7}{10}$ _____

   (4) ____ < ____ < ____

5. 95% .99 $\frac{9}{10}$

   (1) 95% _____

   (2) .99 _____

   (3) $\frac{9}{10}$ _____

   (4) ____ < ____ < ____

6. 1% 1 $\frac{1}{4}$

   (1) 1% _____

   (2) 1 _____

   (3) $\frac{1}{4}$ _____

   (4) ____ < ____ < ____

Change each group of numbers to percents. Then, write the numbers
in order from largest to smallest.

7. $\frac{4}{10}$ $\frac{1}{5}$ $\frac{3}{4}$

   (1) $\frac{4}{10}$ _____

   (2) $\frac{1}{5}$ _____

   (3) $\frac{3}{4}$ _____

   (4) ____ > ____ > ____

8. $\frac{1}{2}$ $\frac{1}{3}$ $\frac{3}{8}$

   (1) $\frac{1}{2}$ _____

   (2) $\frac{1}{3}$ _____

   (3) $\frac{3}{8}$ _____

   (4) ____ > ____ > ____

9. $\frac{3}{5}$ $\frac{3}{4}$ $\frac{2}{8}$

   (1) $\frac{3}{5}$ _____

   (2) $\frac{3}{4}$ _____

   (3) $\frac{2}{8}$ _____

   (4) ____ > ____ > ____

Problem Solving—Percent Problems

The steps you have learned to solve word problems can be used with word problems that deal with percents.

Step 1   Read the problem and underline the key words. These words will usually relate to some mathematics reasoning computation.

Step 2   Make a plan to solve the problem. Ask yourself, Should I add, subtract, multiply, divide, round, or compare? You may have to do more than one of these operations for the same problem.

Step 3   Find the solution. Use your math knowledge to find your answer.

Step 4   Check the answer. Ask yourself, Is the answer reasonable? Did you find what you were asked for?

When you begin with the first step, read the problem carefully. Look for clue words that will give you important information. For example, some problems will provide information about percents.

Examples

A.   Carlos lives in an apartment building that has fireplaces. If 100% of the apartments have fireplaces, what percent of the apartments do not have fireplaces?

Step 1   Determine what **percent** of the apartments do not have fireplaces. The key words are **100%** and **not.**

Step 2   The key words indicate that this is a percent problem.

Step 3   Solve the problem.

     100% means the whole amount. In this problem, it means all the apartments have fireplaces.

     0% means none. So, 0% of the apartments **do not** have fireplaces.

Step 4   Check the answer. One way to make sure the answer makes sense is to recall that percent problems are used to relate different numbers to each other in terms of 100. If 100% means the whole amount or whole unit, then all the apartments must have fireplaces.

B. The supply room is used to store cartons. Twenty-five percent of the cartons hold replacement parts. The rest of the cartons hold tools. What percent of the cartons hold tools?

Step 1 Determine what percent of the cartons hold tools. The key words are **twenty-five percent** and **tools.**

Step 2 The key words indicate that this is a percent problem. You know that a percent problem is asking the relationship of an amount to the whole unit. In this problem, you already know 25% of the cartons hold replacement parts. Now, you must find the percent that is left to hold tools. You will subtract to find the percent of cartons left.

Step 3 Solve the problem.

100% means the whole amount. In this problem, it means all the cartons.

25% means the percent that holds replacement parts.

100% − 25% = 75%

75% of the cartons hold tools.

Step 4 Check the answer. One way to make sure a subtraction is accurate is to add. Does 25% + 75% = 100%? Yes, it does. The answer is reasonable.

Practice

Problem Solving

Solve the following problems.

1. Everyone in the Southside neighborhood wanted to go to the community play. What percent of the people in the neighborhood did not want to go to the play?

2. No one in the band wanted to go bowling after band practice. What percent of the band did not want to go bowling?

3. All the library books were shelved in alphabetical order. What percent of the books were placed in order by size?

4. Every seat in the auditorium was filled. What percent of the auditorium was empty?

5. Sixty percent of the money raised for the school band was earned by the students. The rest was donated by local businesses. What percent was donated by local businesses?

_____

6. Vince did not get a single question wrong on his math test. What percent of the test did Vince get right?

_____

7. Ninety-nine percent of the cars coming off the production line were flawless. What percent had to be corrected?

_____

8. Twenty-two percent of the shirts sold at the store were purchased by women. What percent of the sales were made to men?

_____

9. Forty percent of traffic violations on the highway are committed by people between the ages of 16 and 21. What percent of the traffic violations are committed by people over the age of 21?

_____

10. The Parades family spent 45% of its income on rent and utilities. What percent of their income is left?

_____

11. The word processing program at the local night school uses 16% of the computer memory. What percent of the computer memory is available for the other programs?

_____

12. Eighty-five percent of the movie tickets were sold to adults. The rest were sold to children. What percent of the tickets were sold to children?

_____

13. Sixty-five percent of the people attending the picnic participated in the three-mile run. What percent did not participate?

_____

14. Two percent of the students attending the lecture were late. What percent of the students arrived on time?

_____

15. Thirty-two percent of Gilda's income comes from tips. The rest is earned as salary. What percent is earned as salary?

_____

The large triangle is divided into 100 parts. Use the triangle to answer questions 1–3.

1. What percent of the triangle is shaded? _____

2. What percent of the triangle is not shaded? _____

3. What percent of the triangle has been divided into smaller triangles? _____

Fill in the chart by changing the given percent, decimal, or fraction to other forms.

	Fraction	Decimal	Percent
4.	_____	_____	95%
5.	_____	1.00	_____
6.	$\frac{1}{4}$	_____	_____
7.	_____	_____	280%
8.	_____	.006	_____
9.	$\frac{2}{5}$	_____	_____
10.	_____	_____	1.6%
11.	$\frac{1}{2}$	_____	_____
12.	_____	.16	_____

Write in order from the largest to the smallest.

13. .83, $\frac{4}{5}$, 8.4% 14. .11%, .5, $\frac{1}{5}$ 15. $\frac{1}{2}$, .55%, .55

——— , ——— , ———   ——— , ——— , ———   ——— , ——— , ———

16. $\frac{1}{3}$, 33%, .35 17. 63%, $\frac{6}{10}$, 6.3 18. 4.2%, $4\frac{1}{2}$, 4.55

——— , ——— , ———   ——— , ——— , ———   ——— , ——— , ———

─────────────────── **Problem Solving** ───────────────────

Solve the following problems.

19. The air we breathe is $\frac{1}{5}$ oxygen. What percent is oxygen? _____

20. Nitrogen is .78 of the air. What percent is nitrogen? _____

21. .93% of the air is argon. Change the percent to a decimal. _____

22. After the weather forecaster predicted a major snowstorm, none of the grocery stores in town had bread. What percent of the grocery stores were out of bread? _____

23. Five percent of the fresh vegetables ordered for the Castillo Restaurant were damaged by frost. What percent was usable? _____

24. Eleven percent of the students attending night classes saw the ad in the newspaper. The remaining students received a flyer in the mail. What percent received the flyer? _____

25. The weather forecaster has been correct 87% of the time. What percent of the time has he been incorrect? _____

26. Seventy-seven percent of the voters approved the school tax referendum. What percent did not approve it? _____

3

Solving Percent Problems

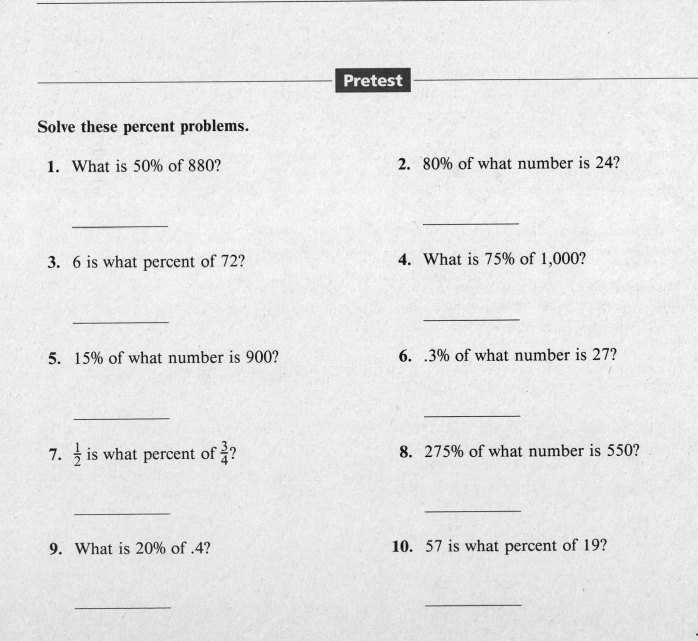

Pretest

Solve these percent problems.

1. What is 50% of 880?

2. 80% of what number is 24?

3. 6 is what percent of 72?

4. What is 75% of 1,000?

5. 15% of what number is 900?

6. .3% of what number is 27?

7. $\frac{1}{2}$ is what percent of $\frac{3}{4}$?

8. 275% of what number is 550?

9. What is 20% of .4?

10. 57 is what percent of 19?

Problem Solving

Solve the following problems.

11. Juanita and Juan want to remodel their kitchen. They plan to buy a freezer. The freezer usually sells for $550. It now costs $429. The sale price is what percent of the original price?

_____

12. Juanita and Juan are also looking for a new stove. The original price was $400. With 25% off this price, how much will they save?

_____

13. The sale price on a microwave oven is $395. This is 79% of the original price. Before the sale, what was the cost of the oven?

_____

14. A dishwasher was $500. It is now marked down 25%. What is the savings amount?

_____

15. The sales tax on all purchases is 7%. The total purchases for the kitchen are $1,499. How much sales tax will Juanita and Juan pay?

_____

Recognizing Percent Problems

Basic percent problems are written in one of these two forms:

50% of 300 is 150.   **or**   75 is 30% of 250.

All percent problems are made up of four elements:
The **whole,** the **part,** the **percent,** and **100.**

> The **whole** is the number value after the word **of.**
> The **part** is listed before or just after the word **is.**
> The **percent** always has a percent sign **(%).**
> All percents are based on **100.**
> The **percent** of a **whole** is the **part.**

Examples

A.   50%   of   300   is   150.
         ↑   ↑   ↑
   percent of **whole** = **part**

The percent is 50 because it has a percent sign.
The whole is 300 because it is the number after the word **of.**
The part is 150 because it is after the word **is.**

B.   75   is   30%   of   250
         ↑   ↑   ↑
   part = **percent** of **whole**

The percent is 30 because it has a percent sign.
The whole is 250 because it is the number after the word **of.**
The part is 75 because it is before the word **is.**

Now organize these problems into grids.

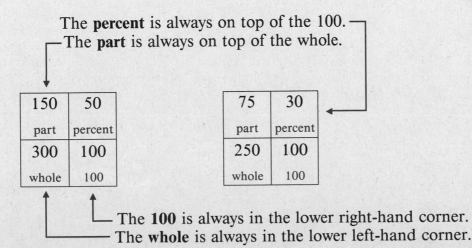

The **percent** is always on top of the 100.
The **part** is always on top of the whole.

150	50
part	percent
300	100
whole	100

75	30
part	percent
250	100
whole	100

The **100** is always in the lower right-hand corner.
The **whole** is always in the lower left-hand corner.

For each of the following percent problems, place the whole, the part, the percent, and 100 on the grid.

1. 82 is 25% of 328.

part	percent
whole	100

2. 25% of 200 is 50.

part	percent
whole	100

3. 133 is 20% of 665.

part	percent
whole	100

4. 19 is 2% of 950.

part	percent
whole	100

5. 6 is $33\frac{1}{3}$% of 18.

part	percent
whole	100

6. 1.5% of 100 is 1.5.

part	percent
whole	100

7. 350% of 24 is 84.

part	percent
whole	100

8. 5 is $\frac{1}{2}$% of 1,000.

part	percent
whole	100

9. 22 is 200% of 11.

part	percent
whole	100

10. 75% of 400 is 300.

part	percent
whole	100

Solving for the Part—Decimal or Fraction Method

When solving for the **part,** you can use one of two methods—(1) the **decimal** method or (2) the **fraction** method. If you have a calculator or if you prefer long division, use the decimal method. If you prefer to use cancellation, use the fraction method. The steps remain the same.

When using the **decimal method,** follow these steps:

Step 1 Identify the parts. Place the known information on a grid.

Step 2 Multiply the shaded diagonals.

Step 3 Divide the answer by the number that is left using long division.

Examples

A. What is 30% of 770?

Step 1 Place the known information on the grid.

?	30
part	percent
770	100
whole	100

Step 2 $770 \times 30 = 23{,}100$

Step 3
$$
\begin{array}{r}
231 \\
100\overline{)23{,}100} \\
\underline{20\,0} \\
3\,10 \\
\underline{3\,00} \\
100 \\
\underline{100}
\end{array}
$$

The answer is 231.
30% of 770 is 231.

B. What is .5% of 300?

Step 1 Place the known information on the grid.

?	.5
part	percent
300	100
whole	100

Step 2
$$
\begin{array}{r}
300 \\
\times\quad .5 \\
\hline
150.0
\end{array}
$$

Step 3
$$
\begin{array}{r}
1.5 \\
100\overline{)150.0} \\
\underline{100} \\
50\,0 \\
\underline{50\,0}
\end{array}
$$

The answer is 1.5.
.5% of 300 is 1.5.

When using the **fraction method,** follow these steps:

Step 1 Identify the parts. Place the known information on a grid.

Step 2 Multiply the diagonals. (This number becomes the numerator.)

Step 3 Divide the answer by the number that is left. (This number becomes the denominator.)

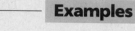 **Examples**

A. What is 45% of 980?

Step 1 Place the known information on the grid.

?	45
part	percent
980	100
whole	100

Step 2 45×980

Step 3 $\dfrac{(45 \times 980)}{100} = \dfrac{882}{2} = 441$

The answer is 441.
441 is 45% of 980.

B. What is 65% of 240?

Step 1 Place the known information on the grid.

?	65
part	percent
240	100
whole	100

Step 2 65×240

Step 3 $\dfrac{(65 \times 240)}{100} = \dfrac{156}{1} = 156$

The answer is 156.
156 is 65% of 240.

Practice

Place the known information on the grid. Solve for the part using the decimal method.

1. What is 24% of 1,080?

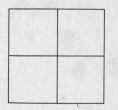

_____

2. What is 99% of 200?

_____

3. What is 75% of 440?

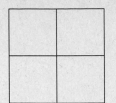

_____

4. What is 23% of 1,800?

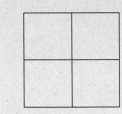

_____

5. What is 10% of 6,300?

_____

6. What is 18% of 500?

_____

Place the known information on the grid. Solve for the part using the fraction method.

7. What is 5% of 275?

_____

8. What is 10% of 10?

_____

9. What is 58% of 440?

_____

10. What is 20% of 70?

_____

11. What is 80% of 80?

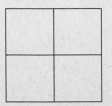

_____

12. What is 75% of 28?

_____

Estimating the Part

An estimate is an intelligent guess about an answer to a problem. Before you begin to solve a percent problem, estimate the answer by rounding off all of the known values to whole numbers. Quickly fill out a grid. Most of the time you should be able to do the math in your head. Make a note of the estimated answer; then actually solve the problem. If the two answers are close, then your answer is probably correct.

Example

Carmen bought a bracelet at Chu Lin's Golden Charms. The cost of the bracelet was $885. If the tax rate was 7.4%, how much was the tax?

Round off each of the known values.

Rounded
Percent   7%
Part   unknown
Whole $900

Actual
Percent   7.4%
Part   unknown
Whole $885

Fill in the grid.

?	7
part	percent
900	100
whole	100

Fill in the grid.

?	7.4
part	percent
885	100
whole	100

Multiply the diagonals.

$900 \times 7 = 6,300$

Divide by the number that is left.

$6,300 \div 100 = 63$

The estimated tax is $63.

Multiply the diagonals.

$885 \times 7.4 = 65,490$

Divide by the number that is left.

$65,490 \div 100 = 65.49$

The actual tax is $65.49.

Estimate the solutions to each part of the following problem, then solve for the actual answers.

Marda works in a restaurant. She is required to add 6.5% sales tax to the price of each dinner. Here is a list of her dinner bills for Saturday. Find the sales tax on each one.

	Estimate	Actual
1. $10.40	_____	_____
2. $32.50	_____	_____
3. $27.00	_____	_____
4. $110.00	_____	_____
5. $96.00	_____	_____

Shopping at a Sale

Read the following conversation between Constanza and J. D. Gonzalez while they were shopping at an electronics store. Find the savings on each item they planned to purchase. Write the answer on the line provided.

1. "The $480 stereo system I wanted is on sale for 25% off. That's a savings of $_____," said J.D.

   _____

2. "You'll need speakers, too. They are marked down 40% from $375. You will save $_____," said Constanza.

   _____

3. "The extra long wires are $5.50 a set. That's 20% off. I'll save $_____," said J.D.

   _____

4. "CDs are 15% off when you buy the stereo. At $10 each, you'll save $_____ on each one," said Constanza.

   _____

5. "If I get this $425 VCR at 25% off, I'll save $_____," said J.D.

   _____

6. "You can buy this giant screen TV. It was originally $1,995 reduced by 24%. You would save $_____, J.D." said Constanza.

   _____

7. "A laser disc player plays movies, pictures, and music. Last week it was $675, but today, I'll give you $33\frac{1}{3}$% off. You'll save $_____," said the salesperson.

"You know, Constanza, we could go broke saving money," said J.D. as he walked out of the store.

Problem Solving—Finding the Part of the Percent

When you take exams such as the GED or an employment test, you will find that most percent problems are presented as word problems. One of the most important things to remember with a percent problem is to simplify it to the sentence: **The percent of the whole is the part.** Word problems are not difficult to master if you follow these steps:

Step 1 Read the problem and simplify it to the sentence: **The percent of the whole is the part.**

Step 2 Make a plan to solve the problem. Part of the plan is to draw and label a grid.

part	percent
whole	100

Step 3 Put the known information in the correct places on the grid.

Step 4 Multiply the diagonals, then divide by the number that is left.

Example

The new furniture the Garcia family wanted to buy was on sale. The original price was $2,400. The sale price of all the furniture was 75% of the original price. What was the sale price?

Step 1 Simplify the problem to the sentence: **The percent (75%) of the whole ($2,400) is the part (sale) price.**
The part or sale price is the unknown. This is what you are to solve for.

Step 2 Draw a grid.

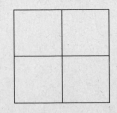

Step 3   Put the known information in the correct places on the grid.

?	75
part	percent
2,400	
whole	100

Step 4   Multiply the diagonals; then divide by the number that is left.

$$\frac{2,400 \times 75}{100}$$

$$\frac{\overset{24}{\cancel{2400}} \times 75}{\underset{1}{\cancel{100}}} = 1,800$$

$1,800 is 75% of $2,400.
The furniture is on sale for $1,800.

Practice

Solve the following problems. Use the grids provided and write your answers on the lines.

1. 32% of the vote has been counted. If 775 people voted, how many votes have been counted so far?

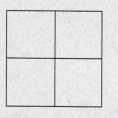

_____

2. 62.5% of the people polled did not agree with the Senate's vote. If 8,000 people were polled, how many did not agree?

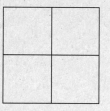

_____

3. 38% of the families in a neighborhood are against having a park nearby. If there are 200 families in the area, how many are against the park?

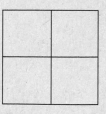

_____

4. CoraSue bought a desk for $55. After she refinished it, she sold it at 250% of its original price. What did she sell it for?

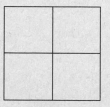

_____

Solving for the Percent—Decimal or Fraction Method

When solving for the **percent,** you can use one of two methods—
(1) the **decimal** method or (2) the **fraction** method.
When using the decimal method, follow these steps:

Step 1 Place the known information on a grid.

Step 2 Multiply the diagonals.

Step 3 Divide the answer by the number that is left.

Step 4 Add the percent sign.

Examples

A. 13 is what percent of 25?

Step 1 Place the known information on a grid.

13	?
part	percent
25	100
whole	100

Step 2 Multiply the diagonals.

$$13 \times 100 = 1,300$$

Step 3 Divide by the number that is left.

$$\begin{array}{r} 52 \\ 25\overline{)1,300} \\ \underline{1\ 25} \\ 50 \\ \underline{50} \end{array}$$

Step 4 Add the percent sign.

52%

13 is 52% of 25.

B. 62 is what percent of 124?

Step 1 Place the known information on a grid.

62	?
part	percent
124	100
whole	100

Step 2 Multiply the diagonals.

$$62 \times 100 = 6,200$$

Step 3 Divide by the number that is left.

$$\begin{array}{r} 50 \\ 124\overline{)6,200} \\ \underline{6\ 20} \\ 00 \\ \underline{00} \end{array}$$

Step 4 Add the percent sign.

50%

62 is 50% of 124.

When using the fraction method, follow these steps:

Step 1 Place the known information on a grid.

Step 2 Multiply the diagonals. (This number becomes the numerator.)

Step 3 Divide the answer by the number that is left. (This number becomes the denominator.)

Step 4 Add the percent sign.

--- **Examples** ---

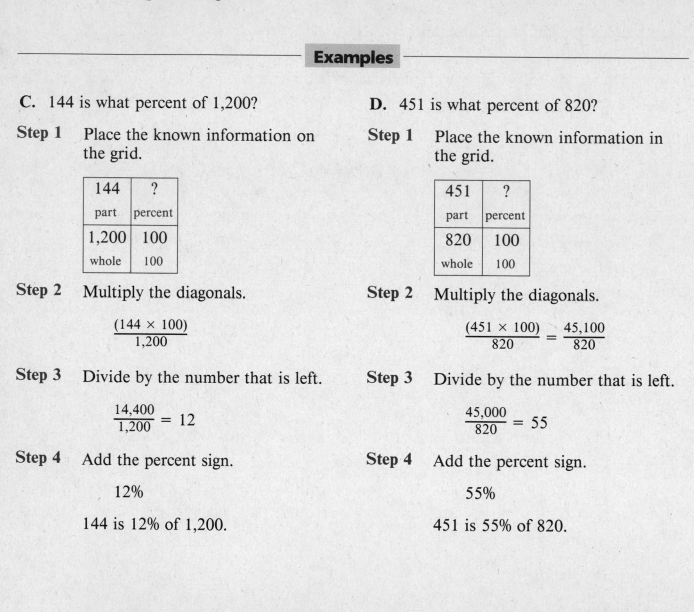

C. 144 is what percent of 1,200?

Step 1 Place the known information on the grid.

144	?
part	percent
1,200	100
whole	100

Step 2 Multiply the diagonals.

$$\frac{(144 \times 100)}{1,200}$$

Step 3 Divide by the number that is left.

$$\frac{14,400}{1,200} = 12$$

Step 4 Add the percent sign.

12%

144 is 12% of 1,200.

D. 451 is what percent of 820?

Step 1 Place the known information in the grid.

451	?
part	percent
820	100
whole	100

Step 2 Multiply the diagonals.

$$\frac{(451 \times 100)}{820} = \frac{45,100}{820}$$

Step 3 Divide by the number that is left.

$$\frac{45,000}{820} = 55$$

Step 4 Add the percent sign.

55%

451 is 55% of 820.

Place the known information on the grid. Solve for the percent using
the decimal method.

1. 84 is what percent of 336?

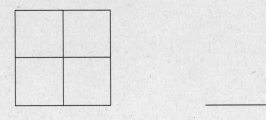

_____

2. 620 is what percent of 3,100?

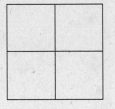

_____

3. 83 is what percent of 332?

_____

4. What percent of 2,880 is 720?

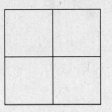

_____

5. 789 is what percent of 789?

_____

6. 96 is what percent of 640?

_____

7. 6 is what percent of 10?

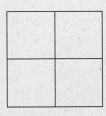

_____

8. 352 is what percent of 640?

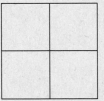

_____

**Place the known information on the grid. Solve for the percent using
the fraction method.**

9. 25 is what percent of 25?

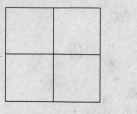

_____

10. 6 is what percent of 24?

_____

11. 17 is what percent of 85?

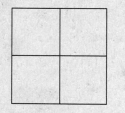

_____

12. What percent of 500 is 50?

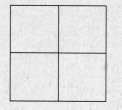

_____

13. 48 is what percent of 12?

_____

14. 49 is what percent of 98?

_____

Estimating the Percent

When estimating the percent, follow these steps:

Step 1 Identify the parts to be placed on a grid.

Step 2 Round off each of the values.

Step 3 Place the known information on the grid and solve.

Example

Edith negotiated the price of a car with a salesperson. The original price of the car was $27,800. She was able to reduce the cost of the car by $1,890.40. Find the percent of estimated savings. Then find the percent of actual savings on the cost of the car.

Step 1 Identify the parts to be placed on the grids.

Percent	unknown
Part	$1,890.40
Whole	$27,800

Step 3 Fill in the grid to find the estimated savings.

2,000	?
part	percent
28,000	100
whole	100

$$\frac{2,000 \times 100}{28,000}$$

$$\frac{\overset{1}{2000} \times \overset{50}{100}}{\underset{\underset{7}{14}}{28000}} = \frac{50}{7} = 7\frac{2}{7}$$

Round the answer to 7%.
The estimated savings was 7%.

Step 2 Round off each of the known values.

Percent	unknown
Part	$2,000
Whole	$28,000

Step 3 Fill in the grid to find the actual savings.

1,890.40	?
part	percent
27,800	100
whole	100

$$1,890.40 \times 100 = 189,040$$

$$189,040 \div 27,800 = 6.8\%$$

The actual savings was 6.8%.

Estimate the solutions to the following problems; then solve for the actual answer.

The Aqualina family pays a total of $4,700 a year for insurance. This amount includes $987 for life insurance; $1,410 for car insurance; $1,692 for insurance on the motorcycle; and $611 for insurance on the old truck. Estimate what percentage of the Aqualina's insurance bill is paid for each of these items. Then calculate the actual percentages.

		Estimate	Actual
1. $987	(life insurance)	_____	_____
2. $1,410	(car insurance)	_____	_____
3. $1,692	(insurance on the motorcycle)	_____	_____
4. $611	(insurance on the old truck)	_____	_____

LIFE SKILL

Percents and Budgeting

Kim and Lu Swanson wondered where all their money went. They decided to figure out what percent of their income they spend on each budget item. Find the answers for the blanks in these problems.

1. "Your net salary is $2,530 a month and mine is $1,256 a month," said Kim. "Our total take-home pay is $_____ a month."

   _____

2. We spend $870.78 a month on rent. That is _____ percent of our income," said Lu.

   _____

3. "The car payments total $795.06 a month," said Kim. "That is _____ percent of our income."

   _____

4. "And gasoline and parking cost $246.09 each month. That's another _____ percent."

   _____

5. Lu added, "We spend $530.04 a month for food. That is _____ percent of our income."

   _____

6. "I enrolled in the major medical insurance program at work. It costs $18.93 a month and is _____ percent of our income," said Kim.

   _____

7. "Can we save $719.34 a month?" asked Kim. "If we save that amount a month, at the end of two years we could put a down payment on a house."

   _____

8. "After we budget these items, we will have $609.76 for clothes, movies, and tickets," said Lu. "That's _____ percent of our income."

   _____

Problem Solving—Finding the Percent

Solving word problems requires patience. Try to think of situations you have encountered that are similar to the ones presented in this lesson. When solving word problems involving percents, remember to follow these steps:

Step 1   Read the problem and simplify it to the sentence: **The percent of the whole is the part.**

Step 2   Make a plan to solve the problem. Part of the plan is to draw and label a grid.

Step 3   Put the known information in the correct places on the grid.

Step 4   Multiply the diagonals; then divide by the number that is left.

Example

At the Skyler Day Care Center, 58 of the 145 children are under three years of age. What percent of the children are under three?

Step 1   Simplify the problem to "The percent (unknown) of the whole (145 children) is the part (58 children)." The percent is the unknown. This is what you will be solving for.

Step 2   Draw a grid.

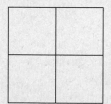

Step 3   Put the known information in the correct places on the grid.

58	?
part	percent
145	100
whole	100

Step 4   Multiply the diagonals; then divide by the number that is
left on the grid.

$$58 \times 100 = 5{,}800$$

$$
\begin{array}{r}
40 \quad \text{(Add the percent sign.)} \\
145\overline{)5{,}800} \\
\underline{5\ 80} \\
00 \\
\underline{00}
\end{array}
$$

58 is 40% of 145.
40% of the children are under three.

Practice

**Solve the following problems. Use the grids provided and write your
answers on the lines.**

1. Everett and Clarice found a used dining
    room set for $240. A new one just like it
    would cost $720. The used furniture
    price is what percent of the price of new
    furniture?

    _____

2. Mario pays $14.40 a month in union
    dues. He earns $1,800 a month. What
    percent of his total pay are his union
    dues?

    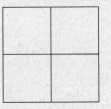

    _____

3. Out of the 51 questions on the test,
    Victor answered 34 correctly. What
    percent of the total is that?

    _____

4. The Johnsons paid $2,320 in real estate
    taxes on their $58,000 townhouse. What
    percentage is their tax rate?

    _____

Solving for the Whole—Decimal or Fraction Method

When solving for the **whole,** you can use one of two methods—
(1) the **decimal** method or (2) the **fraction** method.
When using the decimal method, follow these steps:

Step 1 Place the known information on a grid.

Step 2 Multiply the diagonals.

Step 3 Divide the answer by the number that is left.

Examples

A. 65% of what is 130?

Step 1 Place the known information on the grid.

130	65
part	percent
?	100
whole	100

Step 2 Multiply the diagonals.

$$130 \times 100 = 13,000$$

Step 3 Divide by the number that is left.

```
        200
65)13,000
    13 0
       00
       00
```

The answer is 200.
65% of 200 is 130.

B. 12.5% of what is $50?

Step 1 Place the known information on the grid.

50	12.5
part	percent
?	100
whole	100

Step 2 Multiply the diagonals.

$$50 \times 100 = 5,000$$

Step 3 Divide by the number that is left.

```
          400.00
12.5)5000.00
       500
        00
        00
```

The answer is $400.
12.5% of $400 is $50.

When using the fraction method, follow these steps:

Step 1 Place the known information on a grid.

Step 2 Multiply the diagonals. (This number becomes the numerator.)

Step 3 Divide the answer by the number that is left. (This number becomes the denominator.)

Examples

A. 50% of what is 290?

Step 1 Place the known information on the grid.

290	50
part	percent
?	100
whole	100

Step 2 Multiply the diagonals.

$$290 \times 100$$

Step 3 Divide by the number that is left.

$$\frac{\overset{29}{290} \times \overset{20}{100}}{\underset{\underset{1}{8}}{50}} = 580$$

50% of 580 is 290.

B. 175% of what is 770?

Step 1 Place the known information on the grid.

770	175
part	percent
?	100
whole	100

Step 2 Multiply the diagonals.

$$770 \times 100$$

Step 3 Divide by the number that is left.

$$\frac{\overset{110}{770} \times \overset{4}{100}}{\underset{7}{175}} = 440$$

770 is 175% of 440.

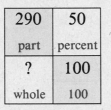

MATH HINT

Remember to cross cancel whenever possible.

Place the known information on the grid. Solve for the whole using the decimal method.

1. 3 is 50% of what?

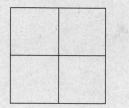

_____

2. 150% of what is 240?

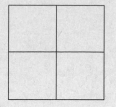

_____

3. 150% of what is 6?

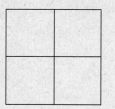

_____

4. 50% of what is 45?

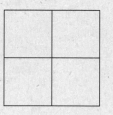

_____

5. 440 is 55% of what?

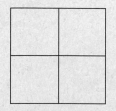

_____

6. 38 is 100% of what?

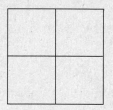

_____

7. 844 is 80% of what?

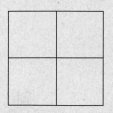

_____

8. 20% of what is 133?

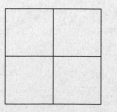

_____

9. 66% of what is 528?

_____

10. 1,350 is 75% of what?

_____

Place the known information on the grid. Solve for the whole using the fraction method.

11. 1% of what is 2?

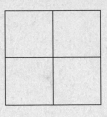

_____

12. 23% of what is 92?

_____

13. 77 is 11% of what?

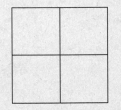

_____

14. 30% of what is 21?

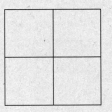

_____

15. 20 is 8% of what?

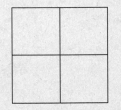

_____

16. 75% of what is 18?

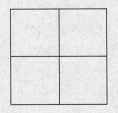

_____

Estimating the Whole

An estimate is an intelligent guess about an answer to a percent problem. Estimate the answer by rounding off the known values in the problem. Quickly fill out a grid and solve for the unknown using easy multiplication and division. Make a note of the estimated answer; then actually solve the problem. If the two answers are close, then your answer is probably correct.

Example

Franco bought a new car. His down payment was 26% of the cost of the car. The down payment was $2,990. What is the cost of the car?

Identify the parts of the grid.
Round off each of the values.

Rounded
Percent   30%
Part   3,000
Whole   Unknown

3,000	30
part	percent
	100
whole	100

$$\frac{3,000 \times 100}{30} = 10,000$$

The estimated cost of the car is $10,000.

Actual
Percent   26%
Part $2,990
Whole   Unknown

2,990	26
part	percent
	100
whole	100

$$\frac{2,990 \times 100}{26} = 11,500$$

The actual cost of the car is $11,500.

> **MATH HINT**
>
> The more you round off the values, the less accurate the answer.

Estimate the solutions to the following problem; then solve for the actual answers.

The following people bought cars. Their down payments were 16% of the cost of the cars. Using the information below, estimate the cost of the car; then find the actual cost.

Name	Down Payment	Estimate Cost	Actual Cost
1. Toni Lugotti	$894.60	_____	_____
2. Frank Mason	$707.80	_____	_____
3. Juanita Alverez	$804.44	_____	_____
4. Susan Murray	$1,054.20	_____	_____

Finding the Original Price

Read the following information about a carpet sale. Answer the questions to help Mai Li determine the best purchase. Find the original price of each of the carpets listed below and place the answer in the blank provided.

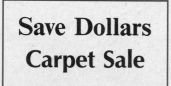

**Save Dollars
Carpet Sale**

Mai Li talked with her friend, Marge, about a carpet sale. Mai Li wanted to know the original price of the carpet before she makes the final purchase.

1.
| Textured Plush
Now $13 a yard. |

The sale price is 65% of the original price. What was the original price?

_____

2.
| Patterned Berber
Now $14.85 a yard. |

This carpet is popular. It is not being reduced as much as some of the others. The price of the carpet is actually 66% of the price before the sale. What did it originally cost?

_____

3.
| Casual Cut and Loop
Now only $30.40 a yard. |

This carpet isn't much of a deal. The sale price is 95% of the original price. What was the original price?

_____

4.

> Special Purchase
> Now only $21 a yard.

This carpet has been marked down twice. It's not selling. The sale price is 42% of the original price. What was the original price?

_____

5.

> Great colors to choose from—Affordable Plush
> Now only $19.80 a yard.

This looks like the best deal in the store. The sale price is 55% of the original price. What was the original price?

_____

6.

> Level Loop with 22 new colors
> Today only $27.60 a yard.

All these colors are really bright. They would be great for children's rooms. The sale price is 75% of the original price. What was the original price?

_____

7.

> Styled yarns with stain release
> Only $31.50 a yard.

This carpet would be good for a room that gets a lot of traffic. The price is 75% of the cost of the carpet before the sale. What did it cost before the sale?

_____

8. Which carpet would you suggest is the best purchase and why? Explain your answer.

_____

_____

_____

Problem Solving—Finding the Whole Number

In previous lessons, you were told that most percent problems are presented as word problems. One of the most important steps when solving a percent word problem is to simplify the word problem. When solving percent problems, follow these steps:

Step 1 Read the problem and simplify it to the sentence: **The percent of the whole is the part.**

Step 2 Make a plan to solve the problem. Part of the plan is to draw a grid.

Step 3 Put the known information in the correct places on the grid.

Step 4 Multiply the diagonals; then divide by the number that is left.

Example

There is a sale at Rogge's Department Store. All the carpet prices are 60% of the original price. If the carpet is now $240, what was the original price?

Step 1 Simplify the problem to "The percent (60%) of the whole (original price/unknown) is the part ($240)."

Step 2 Draw a grid.

Step 3 Put the known information on the grid.

240	60
part	percent
?	100
whole	100

Step 4 Multiply the diagonals; then divide by the number that is left. At this step, you have two methods you can use to solve the problem—the **decimal** method or the **fraction** method.

Decimal Method

$240 \times 100 = 24{,}000$

$$\begin{array}{r} 400 \\ 60\overline{)24{,}000} \\ \underline{24\ 0} \\ 00 \\ \underline{00} \\ 00 \\ \underline{00} \end{array}$$

Fraction Method

$$\frac{240 \times 100}{60}$$

$$\frac{\overset{4}{240} \times 100}{\underset{1}{60}} = 400$$

$240 is 60% of $400.
The original price is $400.00.

Practice

Solve the following problems. Use the grids provided and write your answers on the lines.

1. Luis paid $42 tax on the furniture he bought. If the sales tax is 7%, what was the cost of the furniture?

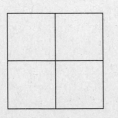

_____

2. The Chins paid $278 in real estate taxes on their home. The tax is .5% of the value of the home. What is their home worth?

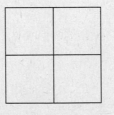

_____

3. Max received a $6.75 tip. It was 15% of the cost of the meal. What was the cost of the meal?

_____

4. To buy a house, the Petries need a down payment of 20%. They have saved $25,000. What is the most the Petries can pay for a house?

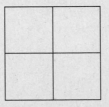

_____

Solving With Mixed Fraction Percents

When solving for the **part** in percent problems involving mixed fractions, follow these steps:

Step 1 Put the known information on a grid.

Step 2 Multiply the diagonals. Change mixed fractions to improper fractions. Use cancellation, if possible.

Step 3 Divide by the number that is left.

Step 4 Reduce answer to lowest terms.

Example

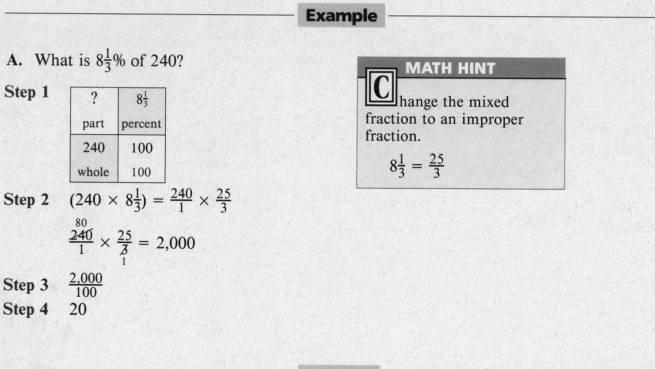

A. What is $8\frac{1}{3}\%$ of 240?

Step 1

?	$8\frac{1}{3}$
part	percent
240	100
whole	100

MATH HINT

$\boxed{C}$hange the mixed fraction to an improper fraction.

$$8\frac{1}{3} = \frac{25}{3}$$

Step 2 $(240 \times 8\frac{1}{3}) = \frac{240}{1} \times \frac{25}{3}$

$$\frac{\overset{80}{\cancel{240}}}{1} \times \frac{25}{\underset{1}{\cancel{3}}} = 2{,}000$$

Step 3 $\frac{2{,}000}{100}$

Step 4 20

Practice
---

1. What is $83\frac{1}{3}\%$ of 54,000?

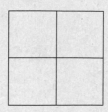

2. What is $\frac{1}{4}\%$ of 440?

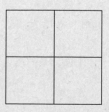

3. What is $66\frac{2}{3}\%$ of 153?

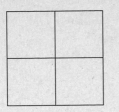 _____

4. $166\frac{2}{3}\%$ of 300 is what?

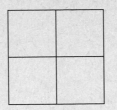 _____

5. What is $33\frac{1}{3}\%$ of 99?

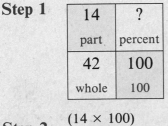 _____

6. $7\frac{1}{2}\%$ of 150 is what?

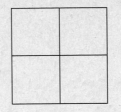 _____

When solving for the **percent** in percent problems involving mixed fractions, follow these steps:

Step 1 Put the known information on a grid.

Step 2 Multiply the diagonals in parentheses. Write the number that is left as the denominator. Use cancellation.

Step 3 Divide. Show the remainder, if any.

Step 4 Rewrite the remainder as a fraction and add the percent sign.

Example

B. 14 is what percent of 42?

Step 1

14	?
part	percent
42	100
whole	100

Step 2 $\dfrac{(14 \times 100)}{42}$

$$\dfrac{\overset{200}{\cancel{1400}}}{\underset{6}{\cancel{42}}} = \dfrac{200}{6}$$

Step 3 $\dfrac{200}{6} = 33$ R2

Step 4 $33\frac{1}{3}\%$

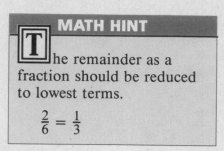

MATH HINT

The remainder as a fraction should be reduced to lowest terms.

$$\frac{2}{6} = \frac{1}{3}$$

Practice

Fill in the grid, then solve for the percent. Write all remainders as fractions.

7.   16 is what percent of 96?

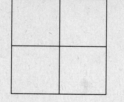

    _____

8.   400 is what percent of 480?

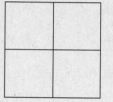

    _____

9.   50 is what percent of 75?

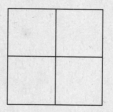

    _____

10.   8 is what percent of 64?

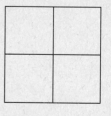

    _____

11.   1 is what percent of 3?

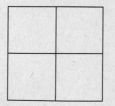

    _____

12.   25 is what percent of 30?

    _____

When solving for the **whole** in percent problems involving mixed fractions, follow these steps:

Step 1 Put the known information on a grid.

Step 2 Change the mixed numbers to improper fractions.

Step 3 Multiply the diagonals within parentheses. Write the number that is left as the denominator.

Step 4 Multiply.

Step 5 Invert and multiply by the denominator. Use cancellation.

Step 6 Reduce to lowest terms.

Example

C. 180 is $33\frac{1}{3}$% of what?

Step 1

180	$33\frac{1}{3}$
part	percent
?	100
whole	100

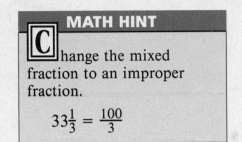

MATH HINT

C hange the mixed fraction to an improper fraction.

$$33\frac{1}{3} = \frac{100}{3}$$

Step 2 $33\frac{1}{3} = \frac{100}{3}$

Step 3 $\dfrac{(180 \times 100)}{\frac{100}{3}}$

Step 4 $\dfrac{18{,}000}{\frac{100}{3}}$

Step 5 $\dfrac{\overset{180}{\cancel{18{,}000}}}{1} \times \dfrac{3}{\underset{1}{\cancel{100}}} = \dfrac{540}{1}$

Step 6 540

Place the known information on the grid. Then solve these problems.

13. $16\frac{2}{3}$% of 216 is what?

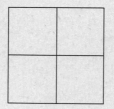

_____

14. $83\frac{1}{3}$% of what is 305?

_____

15. $33\frac{1}{3}$% of what is 9?

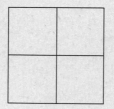

_____

16. $37\frac{1}{2}$% of 384 is what?

_____

17. 81 is $12\frac{1}{2}$% of what?

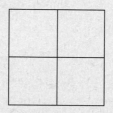

_____

18. $66\frac{2}{3}$% of 69 is what?

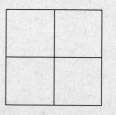

_____

19. 40 is what percent of 64?

_____

20. 105 is what percent of 315?

_____

21. $62\frac{1}{2}$% of what is 145?

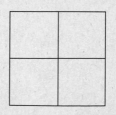

_____

22. $66\frac{2}{3}$% of 513 is what?

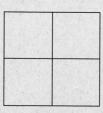

_____

Solving With Fractions

When solving for the **part** in percent problems involving fractions, follow these steps:

Step 1 Put the known information on a grid.

Step 2 Write the diagonals within parentheses.

Step 3 Write the number that is left as the denominator.

Step 4 Multiply within parentheses. Use cancellation.

Step 5 Reduce the new fraction to lowest terms.

Example

A. What is 49% of $\frac{4}{7}$?

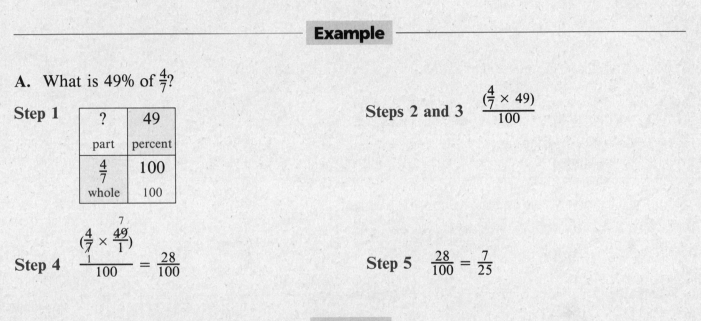

Step 1

?	49
part	percent
$\frac{4}{7}$	100
whole	100

Steps 2 and 3 $\dfrac{(\frac{4}{7} \times 49)}{100}$

Step 4 $\dfrac{(\frac{4}{7} \times \frac{\overset{7}{\cancel{49}}}{1})}{\underset{1}{100}} = \dfrac{28}{100}$

Step 5 $\dfrac{28}{100} = \dfrac{7}{25}$

Practice

Place the known information on the grids. Solve for the unknown parts.

1. What is 50% of $\frac{4}{9}$?

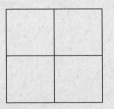

_____

2. What is 150% of $\frac{1}{2}$?

_____

3. What is 40% of $\frac{5}{6}$?

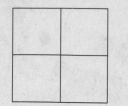

_____

4. What is 20% of $\frac{9}{10}$?

_____

5. What is 50% of $\frac{2}{3}$?

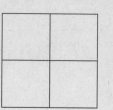

_____

6. What is 75% of $\frac{2}{3}$?

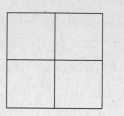

_____

When solving for the **percent** in percent problems involving fractions, follow these steps:

Step 1   Put the known information on a grid.

Step 2   Write the diagonals in parentheses as a numerator.
Multiply the diagonals within the parentheses. Cancel when necessary.

Step 3   Divide by the number remaining on the grid.

Step 4   Simplify the answer and add the percent sign.

Example

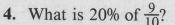

B. $\frac{1}{5}$ is what percent of 2?

Step 1

$\frac{1}{5}$	?
part	percent
2	100
whole	100

Step 2 $\dfrac{(\frac{1}{5} \times 100)}{2} = \dfrac{(\frac{1}{5} \times \frac{\overset{20}{\cancel{100}}}{1})}{2} = \dfrac{20}{2}$

Step 3 $\dfrac{20}{2} = 10$

Step 4   10%

112

Fill in the grid; then solve for the percent.

7. $\frac{4}{5}$ is what percent of $\frac{1}{5}$?

8. $\frac{1}{20}$ is what percent of $\frac{1}{5}$?

9. What percent is $\frac{1}{2}$ of $\frac{1}{10}$?

10. $\frac{2}{5}$ is what percent of $\frac{1}{5}$?

11. $\frac{1}{2}$ is what percent of $\frac{1}{4}$?

12. $\frac{1}{3}$ is what percent of $\frac{2}{3}$?

When solving for the **whole** in percent problems involving fractions, follow these steps:

Step 1   Put the known information on a grid.

Step 2   Write the diagonals in parentheses as a numerator.
Multiply the diagonals within the parentheses. Cancel when necessary.

Step 3   Divide by the number.

Step 4   Simplify the answer, if necessary.

C. 50% of what is $\frac{1}{4}$?

Step 1

$\frac{1}{4}$	50%
part	percent
?	100
whole	100

Step 2 $\dfrac{(\overset{25}{\cancel{100}} \times \frac{1}{4})}{\underset{1}{50}} = \dfrac{25}{50}$

Step 3 $\dfrac{25}{50} = \dfrac{1}{2}$

Fill in the grids; then solve for the whole.

13. $66\frac{2}{3}\%$ of what is $\frac{1}{2}$?

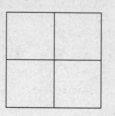

_____

14. 30% of what is $\frac{1}{2}$?

_____

15. 25% of what is $\frac{1}{5}$?

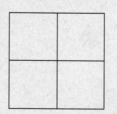

_____

16. 75% of what is $\frac{2}{3}$?

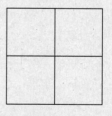

_____

17. 90% of what is $\frac{1}{10}$?

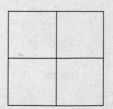

_____

18. $33\frac{1}{3}\%$ of what is $\frac{1}{3}$?

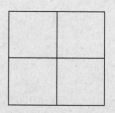

_____

Place the known information on the grids. Solve the following problems. Express all remainders as fractions.

19. $\frac{1}{3}$ is what percent of $\frac{2}{3}$?

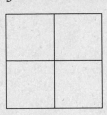

_____

20. $\frac{1}{5}$ is 25% of what?

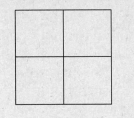

_____

21. $\frac{1}{9}$ is what percent of $\frac{5}{9}$?

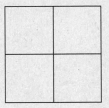

_____

22. What is 25% of 2?

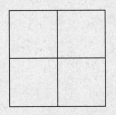

_____

23. $\frac{3}{8}$ is 50% of what?

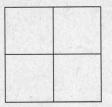

_____

24. $\frac{2}{7}$ is what percent of $\frac{6}{7}$?

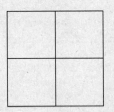

_____

25. $\frac{1}{2}$ is 60% of what?

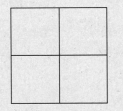

_____

26. $\frac{2}{3}$ is what percent of $\frac{8}{15}$?

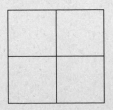

_____

27. 25% of $\frac{1}{5}$ is what?

_____

28. $\frac{3}{4}$ is 150% of what?

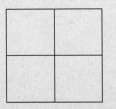

_____

Percent and Patterns

Percent problems relate different numbers to each other in terms of **100**. Remember, a percent problem is made up of the **part**, the **whole**, the **percent**, and **100**. As you study percent problems, you will begin to see some patterns appear.

Example

A. Study the percent problems that follow. Remember, percent of the whole equals the part.

Percent	Whole	Part	Percent	Whole	Part
10% of 500 is		50.	60% of 500 is		300.
20% of 500 is		100.	70% of 500 is		350.
30% of 500 is		150.	80% of 500 is		400.
40% of 500 is		200.	90% of 500 is		450.
50% of 500 is		250.	100% of 500 is		500.

1. How do the percents change?

They increase by tens. This is a pattern of 10.

2. How do the parts change?

They increase by increments of 50. This is also a pattern— a pattern for the part. The pattern is 50.

3. What is the relationship between 10% of 500 and the pattern for the part?

They are the same.
10% of 500 is 50. The part increases by 50.

This is the beginning of a pattern you can use to estimate the percent of a number.

B. What is 70% of 500?
Based on the established pattern, if you multiply 10% by 7 to get 70%, then multiply 50 by 7 to get 350.

$$50 \times 7 = 350$$
70% of 500 = 350.

Use the grid method.

?	70
part	percent
500	100
whole	100

$$\frac{70 \times \overset{5}{\cancel{500}}}{\underset{1}{\cancel{100}}} = 350$$

70% of 500 is 350.

Another way to show this pattern is to draw a table like this:

10%	20%	30%	40%	50%	60%	70%	80%	90%	100%
50	100	150	200	250	300	350	400	450	500

500 is the whole and it is placed under 100%.
50 is 10% of 500, so 50 is placed under the 10%.
250 is 50% of 500, so 250 is placed under the 50%.

Practice

Look for the pattern in the percent problems below. Then answer the questions by referring to the table.

10% of 40 is 4.   50% of 40 is 20.   90% of 40 is 36.
20% of 40 is 8.   60% of 40 is 24.   100% of 40 is 40.
30% of 40 is 12.   70% of 40 is 28.
40% of 40 is 16.   80% of 40 is 32.

10%	20%	30%	40%	50%	60%	70%	80%	90%	100%
4	8	12	16	20	24	28	32	36	40

1. How do the percents change?

2. How does the part change?

3. What is 100% of 40? _____

4. What is 10% of 40? _____

5. What is 50% of 40? _____

6. What is 80% of 40? _____

7. What is the relationship between 10% of 40 and the pattern for increasing the part?

Look at the table below. Fill in the missing information.

10%	20%	30%	40%	50%	60%	70%	80%	90%	100%
24	48			120					240

Complete the pattern started above.

8. 10% of 240 is _____24_____.   9. 60% of 240 is _____.

10. 20% of 240 is _____.   11. 70% of 240 is _____.

12. 30% of 240 is _____.   13. 80% of 240 is _____.

14. 40% of 240 is _____.   15. 90% of 240 is _____.

16. 50% of 240 is _____.   17. 100% of 240 is _____.

Look at the table below. Fill in the missing information.

10%	20%	30%	40%	50%	60%	70%	80%	90%	100%
15	30			75					150

Complete the pattern started above.

18. 10% of 150 is _____15_____.   19. 60% of 150 is _____.

20. 20% of 150 is _____.   21. 70% of 150 is _____.

22. 30% of 150 is _____.   23. 80% of 150 is _____.

24. 40% of 150 is _____.   25. 90% of 150 is _____.

26. 50% of 150 is _____.   27. 100% of 150 is _____.

Look at the table below. Fill in the missing information.

10%	20%	30%	40%	50%	60%	70%	80%	90%	100%
55									

Complete the pattern for the parts started above by first finding the whole, then calculating the rest of the parts.

28. _____55_____ is 10% of 550.

29. _____ is 60% of 550.

30. _____ is 20% of 550.

31. _____ is 70% of 550.

32. _____ is 30% of 550.

33. _____ is 80% of 550.

34. _____ is 40% of 550.

35. _____ is 90% of 550.

36. _____ is 50% of 550.

37. _____ is 100% of 550.

Basic Percent Problems

As you have learned in previous lessons, percent problems consist of four elements:

1. The part
2. The whole
3. The percent
4. 100

> **MATH HINT**
>
> **T**he value after **of** is always the whole.
> The percent always has a **percent sign.**
> The part is just before or after the word **is.**

Basic percent problems are written in one of two forms:

50% of 500 is 250.   **or**   250 is 50% of 500.

50%   of   500   is   250.
↓   ↓   ↓   ↓   ↓
The percent of the whole equals the part.

250   is   50%   of   500.
↓   ↓   ↓   ↓   ↓
The part equals a percent of the whole.

The part can be written first or last.

Remember, in a percent problem, one of the first three elements is missing.

Examples

A. 83% of what is 332?
   The percent is 83 because it has a **percent sign.**
   The word after **of** is **what,** so the whole is missing.
   The number just after **is** is 332, so it is the **part.**
   Place the numbers in the grid.

332	83
part	percent
?	
whole	100

Multiply the diagonals.
$332 \times 100 = 33,200$

Divide by the number that is left.

$83\overline{)33,200} = 400$

83% of 400 is 332.

B. What is 83% of 400?
The percent is 83 because it has a **percent sign.**
The word after **of** is 400.
The number just before **is** is the part and is missing.

83	
part	percent
400	
whole	100

Multiply the diagonals.
$83 \times 400 = 33,200$

Divide by the number that
is left.

$100\overline{)33,200}$

332 is 83% of 400.

C. 332 is what percent of 400?
The percent is missing.
The word after **of** is 400.
The number just before **is** is 332.

332	
part	percent
400	
whole	100

Multiply the diagonals.
$332 \times 100 = 33,200$

Divide by the number that
is left.

$400\overline{)33,200}$

332 is 83% of 400.

--- **Practice** ---

Write the answer to the questions on the lines provided.

1. How can you recognize the number
   that is the whole?

2. How can you recognize the value that
   is the percent?

3. What are two ways to write the percent problem:

Seventy-five percent of 400 is 300.

_____

_____

4. Once you put the numbers in the grid, how can you tell which two numbers should be multiplied?

_____

5. Once you put the numbers in the grid, how can you tell which number is the divisor?

_____

6. Do the known diagonals always slant in the same direction?

_____

Use the grid method to solve the following problems.

7. What is 40% of 164?   _____

8. 50% of 188 is what?   _____

9. 90 is 75% of what?   _____

10. 63% of what is 315?   _____

11. 44 is what percent of 132?   _____

12. 25 is what percent of 300?   _____

13. What is 85% of 600?   _____

14. 42% of what is 882?   _____

15. 5.5% of what is 32.01?   _____

16. What is 20% of 15?   _____

17. 3 is what percent of 6? _____

18. 85 is 50% of what? _____

19. What is 50% of 18? _____

20. 9 is what percent of 36? _____

21. 16% of what is 12? _____

22. 17% of what is 952? _____

23. 8,396 is what percent of 8,396? _____

24. 25 is what percent of 125? _____

25. What percent of 500 is 50? _____

26. 12 is 5% of what? _____

27. What is 66% of 80? _____

28. What percent is 6.3 of 126? _____

29. 90% of what is 270? _____

30. What is 86% of 3,300? _____

Problem Solving Hints

Solutions for percent problems can be written in a variety of ways.
The **fraction method** works well when solving percent problems with
a paper and a pencil. The **decimal method** is quick when you have a
calculator. On some exams, however, only the operational steps to
solve the problem are given. The steps are usually presented in a
form which requires you to know the **order of operations.** When
taking math exams, follow these steps to solve percent problems:

Step 1   Use a grid.

Step 2   Look at your solution; then compare it to the
choices given in the question.

Step 3   Then do one of two things:

    a. Find the solution for each of the choices and
compare it to your answer.

or

    b. Think about what you know about equivalent
fractions, decimals, and percents; and select the
correct alternative.

> **MATH HINT**
>
> **F**ollow these operations
> from left to right:
> 1. Do the math inside the
> parentheses.
> 2. Multiply and divide.
> 3. Add and subtract.

Example

65% of 200 is what?

?	65
part	percent
200	100
whole	100

Multiply the diagonals.
Divide by the number that is left.
The fraction solution looks like this:

$$\frac{65 \times \overset{2}{200}}{\underset{1}{100}} = 130$$

Here are other ways to write the solution to the example.

$65 \times 200 \times .01 = 130$ Multiplying by .01 is the same as
dividing by 100.

$.65 \times 200 = 130$ Multiplying 65 by .01 is .65.

Circle the number of the TWO correct ways of writing the operations for solving each problem. Use the grids to help you.

1. 35% of 8,600 is what?

    (1) $\dfrac{.35 \times 8600}{100}$ (2) $.01 \times 35 \times 8,600$

    (3) $.35 \times 8,600$ (4) $\dfrac{35 \times 100}{8,600}$

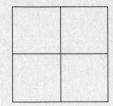

2. 3,150 is what percent of 4,200?

    (1) $\dfrac{3,150 \times 100}{4,200}$ (2) $\dfrac{.01 \times 3,150}{4,200}$

    (3) $\dfrac{100 \times 3,150}{4,200}$ (4) $\dfrac{100 \times 4,200}{3,150}$

3. 80% of 5,300 is what?

    (1) $\dfrac{80(5,300)}{100}$ (2) $\dfrac{.8 \times 5,300}{100}$

    (3) $\dfrac{.8 \times 100}{5,300}$ (4) $.8 \times 5,300$

4. 66% of 700 is what?

    (1) $\dfrac{.66 \times 100}{700}$ (2) $\dfrac{66 \times 700}{100}$

    (3) $.01 \times 66 \times 700$ (4) $\dfrac{.01 \times 66}{100}$

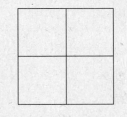

5. What method did you use to select the two correct answers? Explain.

    _____

    _____

Problem Solving—Percent Word Problems

Word problems usually give information in three or four sentences. The numbers for the words in the phrase **"Percent of the whole is the part"** are usually written in different sentences.
To solve percent word problems, follow these steps:

Step 1   Read the problem.

   a. Circle the phrase that means **The percent of the whole is the part.**

   b. Underline the sentence that asks a question. Make sure you understand what the question is asking.

Step 2   Make a plan to solve the problem. Check to determine whether the percent in the phrase matches the percent in the question.

Step 3   **a.** Eliminate unimportant information.

   b. Substitute values for the words in the phrase: **The percent of the whole is the part.**

Step 4   Solve the problem. Use a grid to help you.

Example

Sixty percent of all the students attending classes at the local community college are over the age of 25. The community college enrolls 16,500 students a semester. How many of those students are over the age of 25?

Step 1   Read, circle, and underline.

   (Sixty percent of all the students) attending classes at the local community college are over the age of 25. The community college enrolls 16,500 students a semester. How many of those students are over the age of 25?

Step 2   Make a plan to solve the problem. Check to determine whether the percent in the phrase matches the percent in the question. Sixty percent of all the **students are over the age of twenty-five.** How many of those **students are over the age of twenty-five?** Yes, the percents match. **Students** in the first sentence is repeated as **16,500 students** in the second sentence.

Step 3 Now the phrase is **sixty percent of 16,500 students.**
Remember to change values written as words to numbers.
So, sixty percent becomes 60%.
The phrase now reads: **60% of 16,500 is what?**

Step 4 Solve the problem. Use a grid to help you. Remember to
multiply the diagonals and divide by what is left.

?	60
part	percent
16,500	
whole	100

$$\frac{60 \times \overset{165}{\cancel{16,500}}}{\underset{1}{\cancel{100}}} = 9,900$$

9,900 students are over the age of 25.

Practice

Solve the following problems.

1. Eighty percent of the people enrolled in GED classes said they were planning to continue with vocational training classes. This semester there were 90 people enrolled in classes. How many wanted to take vocational training?

_____

2. Sixty percent of the students taking the GED exam had high enough scores to enter college. For their community, this was a score of 250 and above. If there were 360 students eligible to enter college, how many students took the exam?

_____

3. Changing work schedules made it difficult for 9 of the people enrolled in the class to finish. If there were 45 people in the class, what percent of the class changed work schedules?

_____

4. Fifty percent of the 360 students enrolled in classes wanted to attend the graduation ceremony. How many students wanted to attend the graduation ceremony?

_____

5. Forty percent of the 225 people participating in the college orientation ceremony wanted to bring their teenage children. How many people wanted to bring their children?

_____

6. Sixty-four of the 80 students enrolled in the family literacy classes read to their children every night. What percent of the participants read to their children?

_____

3

Posttest

Solve these percent problems.

1. 341 is what percent of 62? _____

2. 16% of what is 480? _____

3. 57% of 1,000 is what? _____

4. $\frac{1}{2}$ is what percent of $\frac{3}{4}$? _____

5. 8% of what is 150? _____

6. 75% of .5 is what? _____

7. 42 is what percent of 126? _____

8. 35% of what is 140? _____

9. 90% of 470 is what? _____

10. 19 is what percent of 57? _____

For questions 11–15, decide what answers fit in the blanks. Write the answers on the lines provided.

Wayne and Judy Whitebird plan to buy a washer and dryer with their tax refund. Help them figure the cost.

11. "Alvin's Department Store has a washer-dryer pair on sale at 80% of the regular price. The pair now cost $675.00. The original price must have been $_____" said Wayne.

    _____

12. "Don't forget, they charge a delivery and installation fee that is 9% of the sale price," said Judy. "That's an extra $_____."

    _____

13. "6.4% of the sale price is added for sales tax," Wayne said. "That's $_____ tax."

    _____

14. "We'll have to pay $27 to have the old machines hauled away. That's _____% of the sale price," said Judy.

    _____

15. They are selling a five-year warranty for $168.75," said Wayne. "That's _____% of the sale price."

    _____

4

Percent Word Problems

Problem Solving

Solve the following problems.

1. During a special promotion at the auto mall, 37% of the new vehicles sold were trucks; 29% sold were family sedans; 23.5% were vans; and the rest were sports cars. What percent of the total sales were sports cars? _____

2. Curt received a reenlistment bonus of $5,000. He put 20% of it aside for income tax. He invested the rest in equal amounts for a certificate of deposit, a mutual fund, a savings bond, and his savings account. How much did he put into his savings account? _____

3. Bernie earned $2,000 from his second job. He spent 30% of his earnings on new clothes, 15% on a travel TV, and the rest to pay off his credit cards. How much did he owe on his credit cards? _____

4. Fifty percent of the people visiting the aquarium came to see the porpoise show. Fourteen percent viewed the computer exhibit, 26% went to see the rain-forest exhibit, and the rest were there for an educational tour. How many people went on the educational tour?

_____

5. Su-Lin receives a 20% employee discount at the store where she works. How much will she have to pay for a coat that sells for $120?

_____

6. Last year the Flores Bakery sold layer cakes for $12 each. This year the price is $15. What is the percent of increase over last year's price?

_____

7. A computer was on sale for $2,000. Originally it cost $2,500. What was the percent of the discount?

_____

8. The price of fresh fruit is expected to increase by 20%. If the present cost of three pounds of apples is $1.75, what is the expected price after the increase?

_____

9. A shoe sale at Sloan's Department Store reduced shoes by 30%. If a pair of shoes originally cost $65, what is the sale price? If there is a 6% tax on the shoes, what is the final price?

_____

_____

10. Jeral decided to lay patio stones in his yard. The stones normally sell for $4 each. If the stone is on sale for 25% off the original price, how much will 100 stones cost?

_____

Adding and Subtracting Percents

A **whole unit** can be grouped into many parts. No matter how many groups are created, a whole unit or amount is always 100%. Sometimes a small picture can help you better understand a problem.

Example

Twenty-five percent of the people in a community live on farms, 30% live in houses, and the rest live in apartment buildings. What percent live in apartment buildings?

> **MATH HINT**
> The whole unit is 100%.

You can draw a figure to help you see the math problem. Divide the figure into three parts, representing the three different groups. Label one part 25% for the number of people who live on farms; the second part 30% for those who live in houses, and an unknown percent for those living in apartment buildings. Label the whole figure 100%.

55%		?% 100%
25% Farms	30% Houses	Apartment Buildings

The figure helps show that to **find the percent of people living in apartment buildings,** you must:

1. Add the percent of those living in houses to the percent of those living on farms.

25%	Farms
30%	Houses
55%	

2. Subtract the total from 100%.

    $$\begin{array}{r} 100\% \\ -\ 55\% \\ \hline 45\% \end{array}$$

    45% of the people in this community live in apartment buildings.

Solve the following problems. Use the figures provided to help you visualize the problems.

1. There are three shifts at a factory. Forty-three percent of the employees work the day shift. Twenty-two percent work the night shift. What percent work the swing shift?

_____

2. In the new housing development, 15% of the houses will be brick, 44% will be brick and wood, and the rest will be all wood. What percent will be all wood?

_____

3. A delivery company uses vans for 27% of all deliveries. It uses trucks for 33% of its deliveries and cars for 16% of the total deliveries. The company uses bicycles for the rest of its deliveries. What percent of the deliveries are made using bicycles?

_____

4. Forty percent of all sales at a grocery store were made between 4:00 p.m. and 7:00 p.m. Twenty-five percent were made in the morning from 6:00 a.m. to 8:30 a.m. What percent of the sales were made during the rest of the day?

_____

5. Seventy-five percent of Gus' time at work was spent on the phone, calling customers. Thirteen percent of his time was spent typing the orders in the computer. The rest of the time was spent in meetings. What percent of his time was spent in meetings?

_____

Finding the Amount of the Remaining Percent

Some problems give only part of the information needed to solve them. When this occurs, you have to interpret the information given to answer the question. To help you, follow these steps:

Step 1 Identify the parts and match them to their related percents.

Step 2 Find the totals and match them to 100%.

Step 3 Calculate the missing information.

> **MATH HINT**
>
> **A**ny amount can be divided into several parts. Each of the parts can be written as a percent. Added together the parts will always equal the whole amount, or 100%.

Example

Fifty percent of Jules' farm income is based on corn sales. Fifteen percent is based on wheat sales. The rest of Jules' income is earned by selling soy beans. If Jules earns $80,000 a year, how much of his income is based on soy beans?

Create a figure to help you visualize the problem.
The top part of the figure represents the known percentages.
The bottom part of the picture represents the known amounts.

Step 1

50%	15%	?%	100%
Corn	Wheat	Soy Beans	

Step 2 The total percent is 100%. To find the percent for soy beans, you must do the following:

a. Add the percent of corn to the percent of wheat.

$$\begin{array}{r} 50\% \\ +15\% \\ \hline 65\% \end{array}$$

b. Subtract the total from 100%.

$$\begin{array}{r} 100\% \\ -65\% \\ \hline 35\% \end{array}$$

35% of Jules' income is made from soy beans.

Step 3 What is the percent of the whole? Substitute values. Jules' income is $80,000. The simplified problem is:
The income from the sale of soy beans is 35% of $80,000.

Put the information in a grid, multiply the diagonals, and divide by the number that is left.

?	35
part	percent
80,000	100
whole	100

$$\frac{35 \times \overset{800}{\cancel{80,000}}}{\underset{1}{\cancel{100}}} = \$28,000$$

Jules earns $28,000 from the sale of soy beans.

Practice

Read the problems below. Label the missing part, whole, or percent on the figure provided. Shade in the known information. Then, solve the problem. Place the answers on the lines provided.

1. 24,000 people attended the football game. Thirty-five percent of them bought food, 25% bought a souvenir, and the rest bought nothing.

   [ ]

   (1) What percent of the crowd bought food? _____

   (2) How many people bought souvenirs? _____

   (3) How many people bought nothing? _____

2. Josef worked 50 hours last week. He spent 36% of his time arranging goods for shipment, 14% of his time supervising the loading trucks, and 24% of his time doing paper work. The rest of his time was spent writing out the delivery schedule.

   [ ]

   (1) What percent of Josef's time was spent writing out the delivery schedule? _____

   (2) How many hours were spent writing out the delivery schedule? _____

   (3) How many hours were spent arranging goods for shipment? _____

3. Sonja's company makes $60,000 a year catering food. Sixty-five percent of her company's business income is from serving weddings, 15% of its income is from doing evening parties, and the rest of her business' income is from catering for businesses.

(1) What percent of Sonja's business was spent catering for businesses? _____

(2) How much money did she earn from her catering business? _____

(3) How much money did she earn from serving weddings? _____

4. Renee was an aide at a day care center. She spent 20% of her eight-hour day preparing snacks. Thirty-seven percent of her time was playtime. The rest of the day was spent reading stories to the children.

(1) What percent of Renee's time was spent reading stories? _____

(2) How many hours in the eight-hour day were spent preparing snacks? _____

(3) How many hours were spent at playtime? _____

Interpreting Voting Results

Voting results are often given in percents. Complete the table below from the information given. Then answer the questions.

	Total Votes	Votes for Candidate A	Percent of Total	Votes for Candidate B	Percent of Total
Precinct 1	6,500	2,925	_____	3,575	_____
Precinct 2	8,000	_____	50.6%	_____	49.4%
Precinct 3	7,500	_____	$33\frac{1}{3}$%	5,000	_____
Precinct 4	11,250	_____	62%	4,275	_____
Precinct 5	9,400	_____	35%	_____	65%
Precinct 6	10,250	5,125	_____	_____	50%
Grand totals	52,900	24,863	47%	_____	_____

1. What is the total number of votes? _____

2. What is the total number of votes for Candidate B? _____

3. What percent of the vote did Candidate A receive? _____

4. What percent of the vote did Candidate B receive? _____

5. Which candidate won the election? _____

Multiplying and Dividing Percents

In some percent word problems, multiplication clues are used instead of addition and subtraction. Then you are asked to find the remaining amount. Drawing and labeling a figure may help you to understand the problem.

--- **Example** ---

Wing Park was awarded 60% of the insurance money settlement. His two brothers were awarded an equal share of the remainder. If the settlement was $30,000, how much money was awarded to Wing Park's two brothers?

To solve this problem, follow these steps:

Step 1 Draw a figure.

Step 2 Mark off 60% for Wing Park.

Step 3 Then divide the remainder in half. Label it Brother 1 and Brother 2

60%

Wing Park	Brother 1	Brother 2

Since the entire amount is 100%, the two brothers' share is 40%.

$$\begin{array}{r} 100\,\% \\ -\ \ 60\,\% \\ \hline 40\,\% \end{array}$$

The question asks what each brother gets, so 40% must be divided by 2.

$$40\% \div 2 = 20\%$$

Step 4 Now the question can be simplified to the phrase: **20% of $30,000 is what?**

Step 5 Now, use a grid to solve the problem.

?	20
part	percent
30,000	100
whole	100

Multiply the diagonals, and divide by the number that is left.

$$\frac{20 \times 30{,}000}{100} = 6{,}000$$

Each of Wing Park's two brothers received $6,000.

Solve the following problems by labeling the diagram first, then using the grid method to solve the problem.

1. Emilio paid for 25% of the anniversary party for his parents. His three brothers paid an equal share of the rest. If the party cost $500, what did each of the brothers pay?

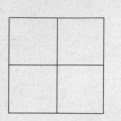

2. John Birdwhistle took a five-day trip. He traveled 40% of his trip on the first day. He traveled an equal distance each of the remaining four days. If the trip was 1,600 miles long, how many miles did he travel on the last day?

3. When the Foster family moved across country, they rented a van. The van held 84% of all their belongings. The rest was divided equally between their two cars. If they moved 5,000 pounds, how much weight was in each car?

4. Ricardo purchased a motorcycle for $4,500. He put a down payment of 60%. He paid the remaining balance of the bike in 15 equal payments. How much were each of the payments?

Selecting Important Information

Many percent problems give more information than is needed. You should delete unimportant information. Double-check the information you use. Make sure it is the same information that you need to solve the problem.

Example

Hank drove 350 miles on Friday, 300 miles on Saturday, and 350 miles on Sunday. What percent of his trip was driven on Saturday?

Step 1 Find the total number of miles driven.

$$\begin{array}{r} 3\,5\,0 \\ 3\,0\,0 \\ +\,3\,5\,0 \\ \hline 1\,,\,0\,0\,0 \text{ miles} \end{array}$$

Step 2 Underline the phrase in the problem that means **the percent of the whole.**
<u>What percent of his trip was driven on Saturday?</u>

Step 3 Use a grid to solve the problem.
The percent is missing.
The part is the mileage driven on Saturday (300 miles).
At this point, the other information (the miles driven on Friday and Sunday) is not needed.

300	?
part	percent
1,000	100
whole	100

Multiply the diagonals and divide by the number that is left.

$$\frac{300 \times \overset{1}{\cancel{100}}}{\underset{10}{\cancel{1000}}} = \frac{300}{10} = 30\%$$

30% of the trip was driven on Saturday.

Solve the following problems. Use the grid method.

1. Wei Yu Li sold a couch for $1,200; a dining room set for $4,400; and a bookcase for $800. The bookcase was what percent of her total sales?

    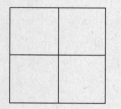

    _____

2. Last year Peggy Nagasawa made $24,000 in commissions. This year she made $27,000 in commissions. Her goal was to earn $36,000 in commissions. What percentage of her goal did she earn this year?

    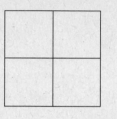

    _____

3. This year 84 of the 120 students in Ms. Greenway's class passed a pretest the first time. Last year only 66 students passed. What percent of this year's students passed the pretest the first time?

    _____

4. The baseball team won 45 out of 60 games during the season. Two games were canceled because of rain. The team's goal was to win 50 games. What percent of all the games played did the team win?

    _____

5. A volunteer committee raised $600 for a middle school band. Last year they raised $750. This year the goal was to raise $800. What percent of the goal has been raised so far?

    _____

6. Josey worked a 10-hour shift on Saturday, a 12-hour shift on Sunday, and a 5-hour shift on Monday. What percent of all his work time was done on Sunday?

    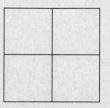

    _____

Percents and Sales Prices

Solve these problems using the information from the advertisement below.

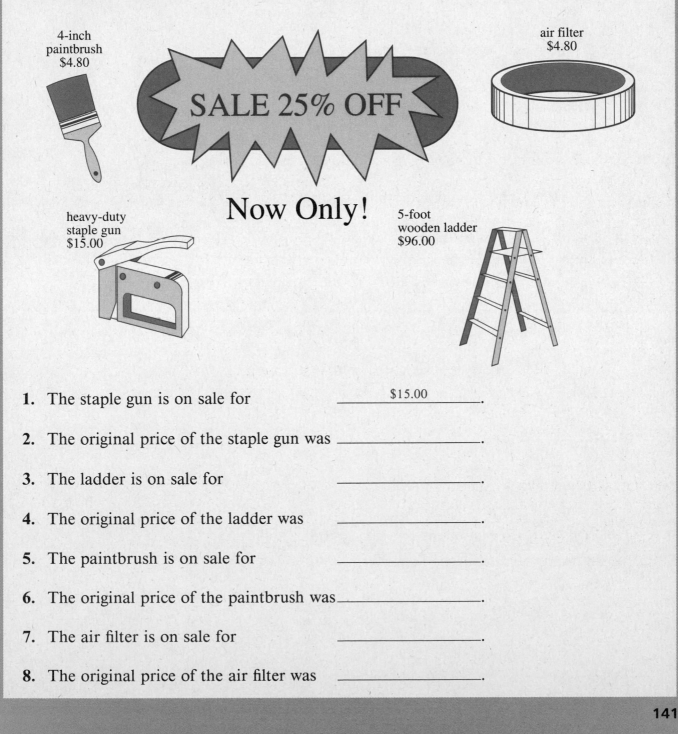

4-inch
paintbrush
$4.80

air filter
$4.80

SALE 25% OFF

Now Only!

heavy-duty
staple gun
$15.00

5-foot
wooden ladder
$96.00

1. The staple gun is on sale for _____ $15.00 _____.

2. The original price of the staple gun was _____.

3. The ladder is on sale for _____.

4. The original price of the ladder was _____.

5. The paintbrush is on sale for _____.

6. The original price of the paintbrush was_____.

7. The air filter is on sale for _____.

8. The original price of the air filter was _____.

Multiple-Step Percent Problems

Many percent problems take two steps to solve. It is very important to think through the problem first; then find the answer to each step of the problem.

Example

Inez works for the ABC Department Store. She gets an employee discount on any clothes she purchases. The tax is added to the discounted price she pays. She buys a pair of slacks which sells for $30. What will she pay if she receives a 20% employee discount and pays a 5% sales tax?

An employee discount of 20% means you should **subtract** 20% from the store price.
Sales tax of 5% means you should **add** 5% to the final price.

You have two percent problems to solve. The decision you have to make is, Which problem should you solve first? First, you must solve for the employee discount and then for the sales tax.

> **MATH HINT**
>
> The sentence **The tax is added to the discounted price she pays** tells you that the tax is calculated after you calculate the employee discount.

Step 1 Calculate the employee discount. Use the grid method.

?	20
part	percent
30	100
whole	100

    a. Multiply the diagonals. $20 \times 30 = 600$

    b. Divide by the number left. $600 \div 100 = 6$

    The employee discount is $6.

Step 2 A discount means you subtract.

$$
\begin{array}{rl}
\$30.00 & \text{cost of slacks} \\
-\quad 6.00 & \text{employee discount} \\
\hline
\$24.00 & \text{price}
\end{array}
$$

Step 3 Find the tax. Use the grid method. Remember to use the new employee price.

?	5
part	percent
24.00	
whole	100

a. Multiply the diagonals. $5 \times 24 = 120$

b. Divide by the number left. $120 \div 100 = 1.20$

The tax is $1.20.

Step 4 Add the tax to the price.

$$
\begin{array}{r}
\$\,2\,4\,.\,0\,0 \\
+\quad 1\,.\,2\,0 \\
\hline
\$\,2\,5\,.\,2\,0
\end{array}
$$

Inez will pay $25.20 for the slacks.

Practice

Decide which percent operation should be done first, then solve the problem.

1. There is a 30% savings on all dishwashers this week. If the original price is $450 and the sales tax is 6%, what is the final cost of the dishwasher?

2. There is a 40% discount on slightly damaged appliances. What is the final cost of a refrigerator originally on sale for $875, if the tax is 4%?

3. There is a 25% discount on all glassware. If a set of glasses was originally marked $125 and the sales tax is 4%, what is the total cost of the glasses?

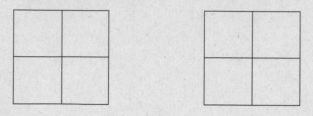

_____

4. Linens are marked 30% off. The original price was $60. If the sales tax is 5%, what is the final price of the linens?

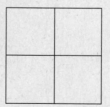 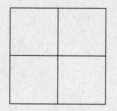

_____

5. A tool shed originally marked $995 is marked 20% off. If the sales tax is 5%, what is the final cost of the tool shed?

_____

6. A clothes dryer originally cost $550. It is marked down 10%. What is the final cost of the dryer with a 7% sales tax?

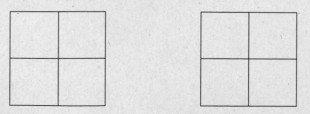

_____

Percent of Increase or Decrease

Until now all the percent problems have been the relationship of a part to the whole.

20% of 50 is 10

Percent of change is used to compare the difference between two numbers. The question usually asks, **What is the percent of increase or decrease between two numbers?**

To solve percent of change questions, follow these steps:

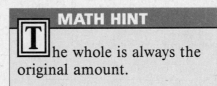

MATH HINT

The whole is always the original amount.

Step 1   Determine the original amount.

Step 2   Determine the new amount.

Step 3   Subtract to find the difference between the new amount and the original amount.

Step 4   The difference becomes the part.

Step 5   Solve for the percent.

Use drawings to help solve percent of increase or decrease problems. In a drawing, the figure representing the original amount and the figure representing 100% are always the same size.

─────────────── **Example** ───────────────

Last year Fred worked 240 days. This year he worked 264 days. What is the percent of increase from last year to this year?

The difference between 240 and 264 is 24.
The question becomes **24 is what percent of 240?**

Use the grid method to solve.

24	?
part	percent
240	100
whole	100

Multiply the diagonals and divide by the number that is left.

$$\frac{\overset{1}{24} \times 100}{\underset{10}{240}} = \frac{100}{10} = 10$$

10% increase from last year.

Another way to solve the problem is to say **264 is what percent of 240?** Then find the difference between the answer and 100%.

264	?
part	percent
240	100
whole	100

$$\frac{264 \times 100}{240} = 110$$

110%

The difference between 110% and 100% is 10%. Fred's work days increased by 10%.

Practice

Solve the following problems. Use a grid to help you.

1. The price of pork changed from $3.50 a pound to $4.20 a pound. By what percent did the price increase? _____

2. The weekly wages of a head chef rose from $700 to $840 in three years. What was the percent of increase? _____

3. Flora and John Sawyer bought a town house for $80,000 seven years ago. They just sold it for $92,000. What was the percent of increase in the value of the house?

_____

4. Henry bought a CD player on sale for $240. It originally sold for $360. What was the percent of decrease?

_____

5. The price of a fabric dropped from $5.50 per yard to $4.29 per yard during a sale. What was the percent of decrease?

_____

6. Portable radios decreased in price from $25 to $15 in the past five years. What is the percent of decrease?

_____

Using Calculators to Solve Percent Problems

To solve percent problems on a simple calculator like the one shown here, all percents must be changed to decimals, and decimals back into percents.

Remember: To change percents to decimals

Step 1   Drop the % sign.

Step 2   Move the decimal point 2 places to the left.

To change decimals to percents

Step 1   Move the decimal point 2 places to the right.

Step 2   Add the % sign.

Readout — Readout

Number key

Function key

Decimal point

Clear

Clear entry

1. What is 15% of 48?
   (Solve for the part.)
   Change 15% to .15.

   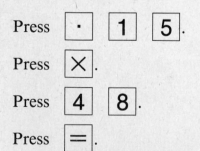

   Read the answer and write it down.
   <u>7.2</u>

   Press [C] before starting the next problem.

2. 362 is 25% of what?
   (Solve for the whole.)
   Change 25% to .25.

   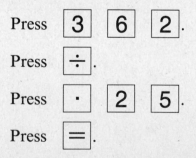

   Read the answer and write it down.
   <u>1,448</u>

   Press [C] before starting the next problem.

3. 3.75 is what percent of 25?
(Solve for the percent.)

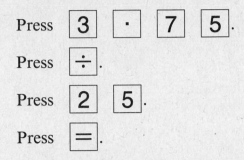

Press [3] [·] [7] [5].

Press [÷].

Press [2] [5].

Press [=].

Read the answer. It is a decimal.
To change the decimal to a percent,
multiply by 100.

Press [×].

Press [1] [0] [0].

Press [=].

Write the answer. Add the % sign.
15%

Work these problems on a calculator.

4. 850 is 20% of what? _____

5. What is 63% of 98? _____

6. What is 152% of 420? _____

7. 8.8 is 110% of what? _____

8. 225 is what percent of 750? _____

Amount of Increase or Decrease

Knowing the percent of change also lets you determine the new amount of increase or decrease.

Example

Culley's Model I camera cost $288 this year. This is 20% more than last year's Model I. What did last year's camera cost?

Look at the figure below:

	100%	20%
Last year's **price**		
This year's **price**		$288

120%

The base year is always 100%. So, 100% appears at the end of the original price.
The increase was 20%. Add 20% to 100%.

$$100\% + 20\% = 120\%$$

120% would appear under $288.

This year's price is 120% of what?
$288 is 120% of what?
Use the grid method to solve the problem.

288	120
part	percent
?	100
whole	100

Multiply the diagonals and divide by the number that is left.

$$\frac{\overset{24}{\cancel{288}} \times \overset{10}{\cancel{100}}}{\underset{\underset{1}{\cancel{12}}}{\cancel{120}}} = 240$$

Last year's camera cost $240.

Solve the following problems. Use the grid method to help you.

1. Last year an average of 11,400 people used the Riverdale transportation system. This year the average number of riders increased by 15%. How many riders did the transportation system average this year?   _____

2. Clancy bought an overcoat last week for $450. This was 25% less than the original price. What was the original price?   _____

3. The Godzik's taxes were increased by 20% over last year. This year their taxes are $1,500. What were last year's taxes?   _____

4. Darla's father bought a gold watch 50 years ago. Today it is worth $1,000. This is an increase of 400%. What was the original price of the watch?   _____

5. James makes $5.52 an hour. This is a 20% increase in salary over last month. What was last month's salary?   _____

6. Werner bought a cycle for $1,925. This represents a 30% decrease in the original cost of the cycle. What was the original cost of the cycle?   _____

7. After the flood, the properties along the river were valued at $24,000 a lot—a 25% drop in the value of the property. What was the price of the property before the flood?   _____

8. The fees for a condominium were $126 a month. This is a 40% increase over fees originally listed. What were the original fees?   _____

LIFE SKILL

Percents and Markup

Store owners buy the goods they sell at wholesale prices (cost). They mark up the prices so that they can make a profit. The amount of markup is often based on a percent of the cost.

Here is how it's done.

Cost + Markup = Selling price

Fill in the missing information on this table of prices.

	Wholesale Price (Cost)	Markup Percent	Markup Amount	Selling Price
1. Washer	$550	_____	$137.50	_____
2. Dryer	$490	37%	_____	_____
3. TV set	_____	75%	$750.00	_____
4. Freezer	$690	_____	$207.00	_____
5. Refrigerator	$450	$33\frac{1}{3}$%	_____	_____
6. Stereo	$375	25%	_____	_____
7. TV set	$800.00	_____	$400	_____
8. Freezer	$700	60%	_____	_____

Problem Solving—Percents

In previous units, you have learned that problem solving is a process which follows certain steps. These steps are listed below:

Step 1 Read the problem and underline the key words. These words will usually relate to some mathematics reasoning computation.

Step 2 Make a plan to solve the problem. Ask yourself, Should I add, subtract, multiply, divide, round, or compare? You may have to do more than one of these operations for the same problem.

Step 3 Find the solution. Use your math knowledge to find your answer.

Step 4 Check the answer. Ask yourself, Is the answer reasonable? Did you find what you were asked for?

Remember, when you are solving percent problems you also can use the **grid method** to help you. Key words in a percent problem are **percent, part,** and **whole.**

--- **Example** ---

Hermano owns a restaurant.
His expenses are as follows:

Fresh meat	$ 900
Fresh vegetables	450
Paper supplies	350
Canned goods	1,000
Fresh breads	300

The paper supplies were what percent of his expenses?

Step 1 Determine what percent of his expenses were his paper supplies. The key word is **percent.**

Step 2 The key word indicates that you are working with a percent problem. You must select the important information. Determine what you know and what you must find out.

Step 3 Solve the problem. A good way to solve this problem is to use the grid method.

    a. Do you know the whole? No, but you can find the whole by adding all the expenses.

Fresh meat	$ 900
Fresh vegetables	450
Paper supplies	350
Canned goods	1,000
Fresh breads	+ 300
Total Expenses	$3,000

    b. Do you know the part? Yes, $350. This was cost of the paper supplies.

    c. Remember the percent will always equal 100%.

You are ready to fill in the grid.

350	?
part	percent
3,000	100
whole	100

Multiply the diagonals and divide by the number that is left.

$$\frac{\overset{35}{\cancel{350}} \times \overset{1}{\cancel{100}}}{\underset{3}{\cancel{\underset{30}{3,000}}}} = \frac{35}{3} = 11\frac{2}{3}\%$$

Paper supplies were $11\frac{2}{3}\%$ of all expenses.

Step 4 Check the answer. You can use a grid again. This time look for the whole. You know the answer should be $3,000. See if you get that answer by using the percent you found.

350	$11\frac{2}{3}$
part	percent
?	100
whole	100

$$\frac{350 \times 100}{11\frac{2}{3}} = \frac{35,000}{\frac{35}{3}}$$

$$35,000 \div \frac{35}{3}$$

$$\frac{\overset{1,000}{\cancel{35,000}}}{1} \times \frac{3}{\underset{1}{\cancel{35}}} = 3,000$$

Your answer is correct.

Solve the following problems. Use the grid method to help you.

1. During a 50-hour work week, Hector spent 38 hours making laser printers, 6 hours inspecting and evaluating them, and 6 hours recording and tracking his progress. What percent of his time was spent making the laser printers?

   _____

2. The trucking schedule was set. Twenty-three percent of the trucks would be carrying produce, 42% would be carrying dry goods, 8% would be carrying furniture. The rest of the trucks would be carrying electronic equipment. If there are 200 trucks, how many are carrying electronic equipment?

   _____

3. The delivery people were responsible for several tasks. Loading their truck took 35% of their time, calling each client took 3% of their time, delivering and setting up the office equipment took 42% of their time. The rest of their time was spent taking payment and filling out the paper work. If they worked a 50-hour week, how much time was spent taking payment and filling out the paper work?

   _____

4. A survey was conducted to determine why people went on vacation. Thirty percent of the people surveyed indicated they went to see family and friends, 26% went on vacation to sightsee, 21% just wanted to stay home, and the rest wanted to go to a resort. If 3,600 people were surveyed, how many wanted to go to a resort?

   _____

5. For every dinner eaten at the restaurant, 45% of the cost of the dinner is the furniture and building, 25% is the cost of personnel, 20% is the cost of the food, and the rest is profit. If a dinner costs $19, how much is profit?

   _____

Figuring Property Tax

Every home has many values.

Current market value is the price for which the home can be sold.

Replacement value is the cost of rebuilding it from scratch.

Assessed value is the value given it by a tax assessor. This is often a percentage of the current market value.

Property tax is based on the assessed value of the home.

From the information given, find the annual property tax for each family's home. The first is worked as an example.

Current Market Value	Assessed Value	Tax Rate	Annual Property Tax
1. Jackson $84,000	$50,400	3.5%	$50,400 × .035 = $1,764
2. Hernandez $103,000	$58,710	3.5%	_____
3. Leong $134,400	$49,728	3.5%	_____
4. Martinelli $99,800	$60,878	2.5%	_____
5. Weinberg $110,800	$61,494	2.5%	_____
6. Littlebird $116,000	$38,860	2.75%	_____
7. McGee $140,000	$88,200	3.0%	_____
8. Olson $178,000	$87,220	2.5%	_____

Posttest

Problem Solving

Solve the following problems. Use grids whenever helpful.

1. A department store earned $60,000 in total sales between Halloween and New Year's Eve. Action toys accounted for 20% of the sales; video games for 25%; dolls and plush animals for 14%. The rest came from the sale of model trains. How much money was earned from model trains?   _____

2. Spike won $2,000 in a state lottery. He put 30% aside for next year's income tax. He divided the rest equally among his wife, each of his two children, and himself. How much money did his wife receive?   _____

3. The Multiplex Theater showed a variety of movies one Friday night. Eighty-two percent of the 2,400 people attending that night went to see a new movie. Twelve percent viewed a forties classic. The remaining people went to see a documentary. What percent went to see the documentary?   _____

4. Two years ago, a large box of breakfast cereal cost $3.60. It now sells for $4.50. By what percent has the cost of the cereal increased?   _____

5. There are 60 attractions at the Lake Amusement Park. Twenty percent are roller coasters; 25% are water rides; 15% are spinning rides; and the rest are children's rides. How many rides are for children?

_____

6. A color TV is on sale for $374. The original price was $440. By what percent was the original price reduced?

_____

7. When the summer travel season begins, the cost of lead free gasoline is expected to increase by 5%. If the gasoline now costs $1.20, how much will it cost when the price increase goes into effect?

_____

8. The new cars on Larry's Lot were damaged in a hail storm. A car originally priced at $18,500 will now be discounted by 12%. How much would the savings be?

_____

9. Tina plans to buy a tent for her yearly camping trip. The original cost was $300. The camping store is having a sale of 6% off all camping items. How much will Tina pay for the tent?

_____

10. Jake is going to panel his living room. The original cost of paneling was $8 a sheet. A spring sale offers 20% off. How much will Jake pay for 10 sheets of paneling?

_____

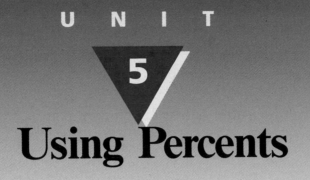

Using Percents

Problem Solving

Find the amount of interest on the following:

1. a 9% medical loan for $3,720 for five years _____

2. a 4.5% home improvement loan for $7,500 for three years _____

3. a $12,450 tuition loan for four years at 7% _____

4. a car loan for $5,220 at 6% for two years _____

5. a furniture loan of $592 at 12% for six months _____

6. a $1,500 savings certificate at 5% a year _____

Read the following information and answer the questions below.

A stereo radio is advertised for $150. It can be paid with a down payment of $50 and six monthly installments of $28.25.

7. What is the total cost of the radio on the installment plan? _____

8. The down payment is what percent of the advertised cost of the radio? _____

9. If a buyer could get a six-month loan of $150 at a 19% annual interest rate for six months, what would the loan cost? _____

10. How much money could be saved by getting a loan instead of paying the installments? _____

11. A washer and dryer are advertised for sale for $1,100. They can be bought with a down payment of $250 and ten monthly installments of $110. _____

    (1) What is the total cost of the washer and dryer if purchased on the installment plan? _____

    (2) The down payment is what percent of the sale price of the washer and dryer? _____

12. If a buyer could get a one-year loan of $1,100 at a 15% annual interest rate for one year, what would be the amount of interest? _____

13. What would be the cost of the loan in Problem 12? _____

14. How much more does the washer and dryer cost when purchased on the installment plan (Problem 11) than they would cost when bought with a loan (Problem 12)? _____

Simple Interest

Simple interest is the cost of using money over a specific length of time.

Money borrowed from a bank, credit union, or a family member is a **loan.** You can get a loan for cars, medical bills, tuition, or a home. A loan for a home is called a **mortgage.**

The amount of money borrowed is the **principal.** The cost of borrowing the money is the **interest.**

Money can be deposited into savings accounts at a bank or a credit union. The amount of money saved is called the **principal.** The amount of money the bank pays into the savings account is also called **interest.** The bank is paying for the use of your money.

To describe the cost of using money, banks and credit unions use the term **rate.**
Rate is the percent of the principal used to calculate the interest.

> **MATH HINT**
>
> **I**nterest is the actual cost of using money over time.
> **Principal** is the amount of money being used.
> **Rate** is the percent of the principal being paid for the use of the money.
> **Time** is expressed in years.

Example

A student decides to take out a loan for tuition. The bank will loan him $2,500 for one year. The bank's interest rate is 9%. The cost of the loan will be $225.

1. **What is the principal?**
   The basic amount of the loan is the principal. $2,500

2. **What is the interest?**
   The cost of the loan is the interest. $225

3. **What is the rate?**
   The rate is the percent of the principal used to
   calculate the interest.   9%

4. **What is the length of the loan?**
   The length of the loan is always written in years.   One year

For each of the problems below, write the principal, the rate, the interest, and the time on the appropriate line.

1. The Alverez family took out a mortgage for $80,000 from a bank. The rate for the loan is 6%. The simple interest for one year is $4,800.

    (1) Principal _____

    (2) Rate _____

    (3) Interest _____

    (4) Length _____

2. The Johansens had a savings account of $5,000. At the end of a year, they had earned $250 in interest. Their interest rate was 5% for a year.

    (1) Principal _____

    (2) Rate _____

    (3) Interest _____

    (4) Length _____

3. Frank Nakai borrowed $750 to purchase a stereo system. The store charged an interest rate of 22%. The interest for one year will be $165.

    (1) Principal _____

    (2) Rate _____

    (3) Interest _____

    (4) Length _____

4. Chu Li bought a savings bond for $1,000 with an interest rate of 5%. At the end of the year, he would earn $50.

    (1) Principal _____

    (2) Rate _____

    (3) Interest _____

    (4) Length _____

5. The MacIntyres bought new furniture. They took out a loan for $3,000. The store's interest rate was 15%. At the end of a year, they would pay an additional $450 for their furniture.

    (1) Principal _____

    (2) Rate _____

    (3) Interest _____

    (4) Length _____

6. Carter borrowed $1,500 from the bank to buy a new car. The rate for the loan is 7% a year. The cost of the loan will be $1,050.

    (1) Principal _____

    (2) Rate _____

    (3) Interest _____

    (4) Length _____

Simple Interest for More Than One Year

The length of time borrowed money is used can vary. Usually money is used for more than one year. To find the interest or cost of borrowing money, follow these steps:

Step 1 Draw a grid with **interest** as the part, **rate** as the percent (per year), and **principal** as the whole.

Step 2 Multiply the diagonals.

Step 3 Divide by the number that is left.

> **MATH HINT**
>
> Simple interest problems are like percent problems except interest problems always include time.
> Part = whole × percent
> Interest = principal × rate × time.

Step 4 Multiply the interest by the time (in years) the money will be used.

Step 5 Write the answer in dollars and cents.

Example

What is the interest on borrowing $250 at a rate of 18% for five years?

Step 1

? interest	18 rate
$250 principal	100 100

Step 2 $\dfrac{18 \times \overset{25}{\cancel{250}}}{\underset{10}{\cancel{100}}} = \dfrac{450}{10}$

Step 3 $\dfrac{450}{10} = 45$ interest

Step 4 45×5 years $= 225$

Step 5 $225 interest for five years at 18% on $250.

Solve the following interest problems. Use the grid and write your answers on the lines provided.

1. What is the interest on $1,025 at a rate of 14% for 6 years?

? interest	rate
principal	100

_____

2. What is the interest on $9,000 at a rate of 9% for 4 years?

? interest	rate
principal	100

_____

3. What is the interest on $2,800 at a rate of 10% for 12 years?

? interest	rate
principal	100

_____

4. What is the interest on $13,000 at a rate of 20% for 13 years?

? interest	rate
principal	100

_____

5. What is the interest on $635 at a rate of 7% for 6 years?

? interest	rate
principal	100

_____

6. What is the interest on $600 at a rate of 9% for 2 years?

? interest	rate
principal	100

_____

7. What is the interest on $2,000 at a rate of 7% interest for 6 years?

? interest	rate
principal	100

_____

8. What is the interest on $450 at a rate of 17% interest for 9 years?

? interest	rate
principal	100

_____

Interest Problems as an Expression

There are many ways to represent an interest problem as an **expression.** On some exams you may take, the answer choices are given as expressions for writing the solutions to a problem rather than the actual solution. When selecting the correct expression for a problem, use the grid method to write out the expression.

Example

Write the expression for the problem:
The interest on $400 at a rate of 20% a year for 3 years is $240.

Substitute the numbers for the words.

? interest	rate
principal	100

$$\text{Interest} = \frac{\text{Rate} \times \text{Principal}}{100} \times \text{Time}$$

$$\frac{20 \times 400}{100} \times \frac{3}{1}$$

$$\text{or } \frac{20 \times (400)}{100} \cdot 3$$

> **MATH HINT**
>
> **M**ultiplying by .01 is the same as dividing by 100. Use order of operations from left to right:
> 1. Math inside parentheses.
> 2. Multiplication and division.
> 3. Addition and subtraction.

Because dividing by 100 is the same as multiplying by .01, this problem could also be written as

$400 \times 20 \times .01 \times 3$

To simplify the answer, you might multiply $20 \times .01$, then write the expression as

$400 \times .2 \times 3$

In order to avoid getting confused, write out the operation using the grid method and compare it to the alternatives.

Circle the numbers of the TWO correct expressions for the following interest problems.

1. A certificate of deposit earning 5% on $2,000 for 10 years.
   (1) $2,000 \times 5 \times .01 \times 10$ (2) $2,000 \times .05 \times 10$
   (3) $\dfrac{2,000 \times .5}{10}$ (4) $\dfrac{2,000 \times .1}{5}$
   (5) $\dfrac{10 \times .05}{2,000}$

2. A federal school loan for $8,000 at 4% interest for 8 years.
   (1) $8,000 \times .04 \times 8$ (2) $\dfrac{8,000 \times 4}{8}$
   (3) $8,000 \times 4 \times .01 \times 8$ (4) $\dfrac{8 \times .04}{8,000}$
   (5) $\dfrac{8,000 \times .04 \times 8}{100}$

3. A home improvement loan for $16,000 at 9% interest for 12 years.
   (1) $\dfrac{16,000 \times .09}{12}$ (2) $\dfrac{16,000 \times 9 \times 12}{100}$
   (3) $\dfrac{16,000 \times 12}{9}$ (4) $16,000 \times .09 \times 12$
   (5) $\dfrac{12 \times .09}{16,000}$

4. A savings account of $2,500 at 3% interest for 2 years.
   (1) $2,500 \times .03 \times 2$ (2) $\dfrac{2,500 \times .03}{2}$
   (3) $\dfrac{2,500 \times 3 \times 2}{100}$ (4) $\dfrac{2 \times .03}{2,500}$
   (5) $2,500 \times .3 \times 2$

5. A certificate of deposit for $5,000 at 7% interest for 20 years.
   (1) $5,000 \times .07 \times 20$ (2) $\dfrac{5,000 \times 7 \times 20}{100}$
   (3) $\dfrac{5,000 \times 20}{.07}$ (4) $\dfrac{5,000 \times .07}{20}$
   (5) $\dfrac{5,000 \times 7}{20}$

Simple Interest Added to the Principal

It is often necessary to find out the total cost of using money. The **sum** of the **interest** and the **principal** is the **total amount of the loan** or **the total amount saved.**

---------- **Example** ----------

What is the total amount of a loan of $48,500 at 12% for 10 years?

Step 1 Draw the grid.

?	12
interest	rate
48,500	100
principal	100

$$\frac{48,500 \times 12}{100} = 5,820$$

Step 2 Multiply by time. $5,820 \times 10 = 58,200$

Step 3 Write the answer in dollars and cents. $58,200

Step 4 Add the interest to the principal.

 $ 5 8 , 2 0 0$
$ 4 8 , 5 0 0$ Principal
$1 0 6 , 7 0 0$ Total amount of the loan

---------- **Practice** ----------

Solve the following interest added to principal problems. Show your work and write your answers on the line provided.

1. What is the total cost of a tuition loan of $9,450 at 10% interest for four years?

?	
interest	rate
principal	100

2. What is the total amount of money saved if $5,000 is left in a savings account at a rate of 6% for seven years?

?	
interest	rate
principal	100

3. What is the total cost of a home improvement loan of $17,500 at 12% interest for three years?

? interest	rate
principal	100

_____

4. What is the total cost of a credit card loan of $2,500 at 18% for six years?

? interest	rate
principal	100

_____

5. What is the total amount of money saved if $12,500 is left in a savings account for ten years at a rate of 4%?

? interest	rate
principal	100

_____

6. What is the total amount of a mortgage for $88,000 at 9% interest rate for 20 years?

? interest	rate
principal	100

_____

7. What is the total amount of a certificate of deposit for $7,800 at 6% for 5 years?

? interest	rate
principal	100

_____

8. What is the total cost of a student loan of $6,500 at 8% for 4 years?

? interest	rate
principal	100

_____

Simple Interest With Decimal Rates

The rates for interest problems may be in decimal form. Use the decimal process for **Solving for the Part** described on page 79.

Example

What is the interest on $850 at a rate of 12.5% for two years?

Step 1   Draw the grid.

? interest	12.5 rate
850 principal	100

Step 2   Multiply the diagonals. $850 × 12.5 = 10,625

Step 3   Divide by the number that is left.   10625 ÷ 100 = 106.25

Step 4   Multiply the interest by the years.   106.25 × 2 = 212.50

Step 5   Write the answer in dollars and cents. $212.50

Practice

Solve the following interest problems. Place the answer on the line provided.

1. What is the interest on $1,400 at a rate of 10.25% for 8 years?

? interest	rate
principal	100

_____

2. What is the interest on $8,234 at a rate of 15.5% for 2 years?

? interest	rate
principal	100

_____

3. What is the interest on $1,025 at a rate of 5.3% for 8 years?

? interest	rate
principal	100

_____

4. What is the interest on $500 at a rate of 3.35% for 2 years?

? interest	rate
principal	100

_____

5. What is the interest on $2,640 at a rate of 10.5% for 10 years?

? interest	rate
principal	100

_____

6. What is the interest on $700 at a rate of 4.75% for 3 years?

? interest	rate
principal	100

_____

7. What is the interest on $540 at a rate of 12.75% for 12 years?

? interest	rate
principal	100

_____

8. What is the interest on $6,950 at a rate of 21.2% for 4 years?

? interest	rate
principal	100

_____

9. What is the interest on $14,250 at a rate of 19.6% for 13 years?

? interest	rate
principal	100

_____

10. What is the interest on $3,560 at a rate of 12.5% for 15 years?

? interest	rate
principal	100

_____

Simple Interest With Fractional Rates

The percentage rates for interest problems may be in fractional form. The process for **Solving the Problem** remains the same. Use the **fraction method** for Solving for the Part described on page 80.

Example

What is the interest on $600 at a rate of $7\frac{1}{2}$% for four years?

Step 1   Draw the grid.

?	$7\frac{1}{2}$
interest	rate
600	
principal	100

Step 2   Change the percent to an improper fraction. $7\frac{1}{2} = \frac{15}{2}$

Step 3   Multiply the diagonals. $\frac{\overset{300}{\cancel{600}}}{1} \times \frac{15}{\underset{1}{\cancel{2}}} = 4,500$

Step 4   Divide by the number that is left. $\frac{4,500}{100} = 45$

Step 5   Multiply by time, reduce, and carry out two decimal places. $\frac{45}{1} \times \frac{4}{1} = \frac{180}{1}$

Step 6   Write the answer in dollars and cents. $180

Practice

Solve the following interest problems. Place the answer on the line provided.

1. What is the interest on $2,800 at a rate of $7\frac{3}{4}$% for eight years?

?	
interest	rate
	100
principal	

_____

2. What is the interest on $1,600 at a rate of $12\frac{1}{4}$% for six years?

?	
interest	rate
	100
principal	

_____

3. What is the interest on $1,000 at a rate of $6\frac{1}{4}$% for two years?

?	
interest	rate
principal	100

_____

4. What is the interest on $550 at a rate of $9\frac{1}{4}$% for fourteen years?

?	
interest	rate
principal	100

_____

5. What is the interest on $169 at a rate of $6\frac{1}{2}$% for two years?

?	
interest	rate
principal	100

_____

6. What is the interest on $252 at a rate of $10\frac{1}{2}$% for four years?

?	
interest	rate
principal	100

_____

7. What is the interest on $1,470 at a rate of $5\frac{1}{4}$% for six years?

?	
interest	rate
principal	100

_____

8. What is the interest on $8,800 at a rate of $3\frac{1}{2}$% interest for 11 years?

?	
interest	rate
principal	100

_____

9. What is the interest on $540 at a rate of $4\frac{1}{2}$% for one year?

?	
interest	rate
principal	100

_____

10. What is the interest on $440 at a rate of $7\frac{1}{4}$% for three years?

?	
interest	rate
principal	100

_____

Changing Time to a Fraction

Borrowed money may be used for any length of time—not just one year at a time.

When time is written as a fraction for interest, it is always written in terms of a year.

The number of months is the numerator and the number of months in a year is the denominator.

$$\frac{\text{part of a year}}{\text{one whole year}}$$

MATH HINT

Number of months = numerator
Number of months in a year $=$ denominator
$\frac{3}{12}$

Examples

A. Write 8 months as a fraction.

$$\frac{8}{12} = \frac{2}{3}$$

B. Write 28 months as a fraction.

$$\frac{28}{12} = \frac{7}{3} = 2\frac{1}{3}$$

C. Write 6 years 3 months as a fraction.

$$6\frac{3}{12} = 6\frac{1}{4}$$

Practice

Write the following terms of a loan as fractions. Reduce to lowest terms.

1. Ten years and three months _____

2. Sixteen months _____

3. Seventy-eight months _____

4. Two months _____

5. Eight months _____

6. Four months _____

7. One year and two months _____

8. Twelve months _____

9. Ten months _____

10. Four years and six months _____

11. Fifteen months _____

12. Sixty-eight months _____

13. Forty-four months _____

14. Ten years and nine months _____

Buying a Car

Doug and Nikki Henderson are finally able to buy a new car. They want to get a bank loan for part of the cost. They can use either First National or Federal Savings. Help them decide which bank has a better loan offer.

1. "We have saved $240 a month for 2 years. We now have a total of $_____ saved," said Nikki.

   _____

2. "We have to put a 30% down payment on the car. The car costs $19,200. That's $_____," said Doug.

   _____

3. "We need a loan for 70% of the cost of the car. That will be $_____," said Nikki.

   _____

4. "The loan for $13,440 from First National is 9% for four years. The interest will be $_____," said Doug.

   _____

5. "At the end of 48 months, we will have paid the bank _____," said Nikki.

_____

6. "The monthly payment will be $_____," said Doug.

_____

7. "We can get a 60-month loan from Federal Savings for $13,440 at a rate of 8%. The total interest on that loan will be $_____," said Nikki.

_____

8. "The total amount of that loan would be $_____," said Doug.

_____

9. "Since we will pay it for over five years, our monthly payment will be $_____," said Nikki.

_____

10. "The loan from _____ will have the lower monthly payment," said Doug.

_____

11. "By the time we pay back the loan, we will pay more interest to _____," said Nikki.

_____

Simple Interest With Time as a Mixed Fraction

To find the interest for years and months, change the number of years or length of the loan to a fraction. Follow the steps in the example.

—————————————————— **Example** ——————————————————

What is the interest on \$7,200 at a rate of 5% for 2 years and 6 months?

Step 1 Draw the grid.

?	5
interest	rate
7,200	100
principal	100

Step 2 Multiply the diagonals and divide by the number that is left.

$$\frac{\overset{72}{\cancel{7,200}} \times 5}{\underset{1}{\cancel{100}}} = 360$$

Step 3 Change the time to a mixed fraction and multiply.

$$2 \text{ years and 6 months} =$$
$$24 + 6 = \frac{30}{12}$$

$$\underset{1}{\frac{30}{\cancel{12}}} \times \frac{\overset{30}{\cancel{360}}}{1} = \frac{900}{1}$$

Step 4 Reduce the answer.

900

Step 5 Write the answer in dollars and cents.

\$900

Solve the following interest problems. Place the answer on the lines provided. Show your work.

1. What is the interest on $3,600 at a rate of 9% for nine years and one month?

?	
interest	rate
principal	100

_____

2. What is the interest on $160 at a rate of 18% for two years and two months?

?	
interest	rate
principal	100

_____

3. What is the interest on $900 at a rate of 8% for one year and three months?

?	
interest	rate
principal	100

_____

4. What is the interest on $2,000 at a rate of 12% for six years and six months?

?	
interest	rate
principal	100

_____

5. What is the interest on $500 at a rate of 12% for twenty-five years and six months?

?	
interest	rate
principal	100

_____

6. What is the interest on $5,500 at a rate of 15% for three years and four months.

?	
interest	rate
principal	100

_____

Simple Interest for Less Than One Year

Many times money is used for less than one year. To find the cost of using money for less than one year, multiply the interest by time written as a fraction of one year. Follow the steps in the example.

Example

What is the interest on $3,900 at a rate of 14% for 6 months?

Step 1 Draw the grid.

? interest	14 rate
3,900 principal	100

Step 2 Multiply the diagonals and divide by the number that is left.

$$\frac{\overset{39}{\cancel{3900}} \times 14}{\underset{1}{\cancel{100}}} = 546$$

> **MATH HINT**
>
> **W**hen doing Step 2, try to reduce the numbers to lowest terms before multiplying and dividing.

Step 3 Represent the months as a fraction of a year.

$$6 \text{ months} = \frac{6}{12} = \frac{1}{2}$$

Step 4 Multiply the interest by time and write the answer in dollars and cents.

$$\frac{546}{1} \times \frac{1}{2} = \$273$$

Solve the following interest problems. Place the answer on the lines provided. Show your work.

1. What is the interest on $8,000 at a rate of 9% interest for 6 months?

? interest	rate
principal	100

_____

2. What is the interest on $48,000 at a rate of 7% interest for 6 months?

? interest	rate
principal	100

_____

3. What is the interest on $10,800 at a rate of 6% interest for 8 months?

? interest	rate
principal	100

_____

4. What is the interest on $16,800 at a rate of 9% interest for 6 months?

? interest	rate
principal	100

_____

5. What is the interest on $660 at a rate of 20% interest for 2 months?

? interest	rate
principal	100

_____

6. What is the interest on $770 at a rate of 12% interest for 2 months?

? interest	rate
principal	100

_____

Savings Account

Money placed in a savings account earns interest. When you open a savings account, you will be given a book to record the money you put into the account (**deposit**), the money you take out (**withdraw**), and the interest you earn. Each time some activity takes place in your account, it is called a transaction.

Look at the picture of the account book below. It has five columns.

Date	The day on which a transaction occurred.
Deposit	Money placed into the account.
Withdrawal	Money taken out of the account.
Interest	The money the bank pays you for the use of your money.
Balance	The total amount of money in your account after a transaction has taken place.

Jose Alvarez   Account Number 1279-99-439-1

Date	Deposit	Withdrawal	Interest	Balance
2/10/--	435			435.00
2/18/--	100			535.00
2/28/--	100			635.00
3/14/--	100			735.00
3/28/--	100			835.00
3/31/--			6.35	841.35
4/6/--		25		816.35
5/15/--	65			881.35
5/30/--		350		531.35
6/30/--			4.32	535.67

Using the information from the entries of the account book, answer the following question.

How many deposits were made in February? _____
Your answer should be 3. The first deposit was on 2/10, the second on 2/18, and the third on 2/28.

Use the account book entries to answer these questions. Place your answers on the blanks provided.

1. How many transactions are shown? _____

2. How many transactions are withdrawals? _____

3. What was the balance on February 18? _____

4. How many transactions are deposits made by Jose? _____

5. What is the total amount of interest earned during the months from February through June? _____

6. What is the total amount of money withdrawn from this account during the months from February through June? _____

7. What was the balance on March 28? _____

8. What was the interest earned shown on March 31? _____

9. What was the amount of the withdrawal on May 30? _____

10. What was the balance on June 30? _____

Compound Interest

In most situations, interest is compounded several times during the year. This means the interest for each part of the year is added to the principal, and that sum is the principal for the next part. Finance charges on some small loans are compounded semiannually. Charge cards usually compound the interest daily.

MATH HINT

Quarterly—every three months or four times a year.
Semiannually—every six months or twice a year.
Daily—every day or 365 times a year.

Example

Read the problem below. Determine how many times the interest on the savings will be compounded over ten years.

The Harris family has a savings account of $500 earning 7% interest a year that is compounded quarterly. If they keep $500 in the account for ten years, how many times during the ten years will the interest be added to the principal.

$$\text{number of years} \times \text{number of times per year}$$
$$10 \text{ years} \times 4 \text{ times per year}$$
$$10 \times 4 = 40$$

Interest will be added to the principal **40** times during the ten-year period.

Practice

Solve the following interest problems. Calculate the number of times the principal must be compounded.

1. A $5,000 savings account earning 4% interest compounded quarterly for three years.

    _____

2. A department store revolving charge account of $500 at a rate of 10%, calculating interest on charges monthly for two years.

    _____

3. A $12,000 loan at 2.9% where interest is calculated daily for three years.

    _____

4. A $2,000 certificate of deposit earning 4% interest compounded semiannually for ten years.

    _____

Finding Compound Interest

When finding interest compounded more than once a year, follow these steps:

Step 1 Determine the number of times interest is added to the principal.

Step 2 Calculate the interest for less than a year.

Step 3 Add the interest to the principal.

Step 4 Repeat Steps 2 and 3 for the length of time the money is borrowed or saved.

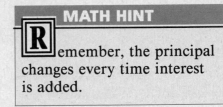

MATH HINT

Remember, the principal changes every time interest is added.

Example

What is the annual interest on $6,000 earning 5% interest compounded semiannually?

Step 1 Determine the number of times the interest must be added to the principal.
Multiply the number of years by the number of times per year.

> One year × two times per year
> 1 × 2

The interest must be calculated and added to the principal two times.

Repeat Steps 2 and 3 **two times.**

Step 2 Calculate the interest on $6,000 for six months.

Step 3 Add the interest to the principal.

First six months

?	5
interest	rate
6,000	
principal	100

Find the interest for 6 months.

$$\frac{5 \times 6{,}000}{100} \times \frac{1}{2} = \frac{300}{2} = \$150$$

Add the interest to the principal.

$6,000 + $150 = $6150

Second six months	? interest	5 rate
	6,500 principal	100 100

Find the interest on the new principal.
$\frac{6{,}150 \times 5}{100} \times \frac{1}{2} = \$153.75$
Add the interest to the new principal.
$\$6{,}150.00 + \$153.75 = \$6{,}303.75$

The interest on $6000 at a rate of 5% compounded semiannually for one year is $303.75.

Practice

Use the grid to find the interest for less than one year, then add it to the principal. Repeat the process for the length of time the money is being used. Put the answer in the blanks provided.

1. What is the annual interest on a savings account with a balance of $6400 at a rate of 10% compounded semiannually?

End of first six months

? interest	rate	Find the interest for six months.	Add the interest to the principal.
principal	100		

Interest

(1) _____

Balance

(2) _____

End of second six months

? interest	rate	Find the interest for six months.	Add the interest to the principal.
principal	100		

Interest

(3) _____

Balance

(4) _____

(5) What is the simple interest for $6,400 at 10% interest for one year.

? interest	rate
principal	100

_____

(6) Which account earns more money, the one using simple interest or the one using compound interest?

_____

(7) Why?

_____

2. What is the interest on $2,560 at 10% compounded quarterly? Round all decimals to nearest whole cent.

End of first three months

? interest	rate	Find the interest for three months.	Add the interest to the principal.
principal	100		

Interest

(1) _____

Balance

(2) _____

End of second three months

? interest	rate	Find the interest for three months.	Add the interest to the principal.
principal	100		

Interest

(3) _____

Balance

(4) _____

End of third three months

? interest	rate	Find the interest for three months.	Add the interest to the principal.
principal	100		

Interest

(5) _____

Balance

(6) _____

End of fourth three months

? interest	rate	Find the interest for three months.	Add the interest to the principal.
principal	100		

Interest

(7) _____

Balance

(8) _____

Choosing an Interest Rate

The Castellanos family has received a $12,000 insurance settlement. They must make some decisions about how they use their money.

Look at the ads below each question. Using what you know about compound interest, circle the letter of the ad that would benefit the Castellanos family the most.

─────────────── **Example** ───────────────

They plan to open a savings account.

A.

> **Guardian Bank**
> 4% interest compounded quarterly

B.

> **Anchor Bank**
> 4% compounded semi-annually

Because the interest at both banks is the same, the difference in earnings will be based on how the interest is compounded. Since Guardian Bank compounds more often, A is a better choice for a savings account.

1. They plan to put $10,000 in a certificate of deposit. Which bank should they select?

   A.

   > **First American**
   > Certificate of Deposit
   > 5% interest
   > compounded quarterly

   B.

   > **First City Bank**
   > Certificate of Deposit
   > 5% interest
   > compounded semi-annually

2. They also intend to purchase a used car. They intend to take out a loan for $4,500 for two years. Which credit union should they select?

A.
> **Farmer's Credit Union**
> car loans, medical loans, machinery loans
> 9.9% interest
> compounded daily

B.
> **Home Savings and Loan**
> car loans
> 9.9%
> compounded monthly

3. They owe $400 to the local department store. Should they stay with the store, or should they put the charges on their credit card?

A.
> **Grant's Department Store**
> Lay-A-Way Policy
> Unpaid balance at the first of the month will be charged 14% compounded monthly.

B.
> **City and Suburb Credit Card**
> Interest as low as 14% compounded daily.

4. The money left from the certificate of deposit will go into a savings account. Which bank should they select?

A.
> **SAVE**
> with
> **Home Savings**
> 5% interest compounded quarterly

B.
> **First Federal Bank**
> 5% interest on all savings accounts
> compounded monthly

Installment Plan Buying

Installment plan buying means paying a certain percent of the purchase price as a **down payment** and agreeing to pay the rest in small amounts each month. The monthly amounts include an interest payment. This type of interest is called a **finance charge**.

Example

Connie bought a living room set on sale for $725. She put down a 10% deposit and paid $55 a month for a year. What was the total amount she paid for the living room set?

$$\text{Deposit} = \frac{725 \times 10}{100} = \$72.50$$

```
  $ 5 5  monthly payment        $  7 2 . 5 0
×    1 2  months              +  6 6 0 . 0 0
  1 1 0                         $ 7 3 2 . 5 0  total cost of furniture
    5 5
  $ 6 6 0
```

Practice

Solve the following problems. Show your work, and write your answers on the lines provided.

1. Elena wants to buy a color TV for $399. The store says she can put 15% down and pay $33.30 monthly for twelve months.

   (1) What is the down payment? _____

   (2) What are the total monthly payments? _____

(3) What is the total cost of the TV? _____

(4) If Elena took out a loan for $399 at 9%, how much would the loan cost for one year? _____

(5) What is the difference in cost between the installment plan and the loan? _____

2. Joseph and his brother, Dave, bought a canoe that cost $1,075. They put 25% down and paid $80 a month for a year.

(1) What was the down payment? _____

(2) What were the total monthly payments? _____

(3) What would the total cost of the canoe be when paid for on the installment plan? _____

(4) If the brothers took out a loan for $1,075 at 10%, how much would the loan cost for one year? _____

(5) What is the difference in cost between the installment plan and the loan? _____

LIFE SKILL

Reading Credit Card Monthly Statements

It is important to know the cost of using a credit card. Take the time to read all the details in your monthly statement.

The statement is the bill from the credit card company. It identifies all the charges and payments made during the month. Any charges not paid in full after the due date are subject to interest or finance charges. The **finance charge** is the cost of borrowing money. The finance charge is calculated on the average daily balance during the month.

Study this portion of a monthly charge account statement.

PREVIOUS BALANCE	AVERAGE DAILY BALANCE	FINANCE CHARGE	PURCHASES	PAYMENTS & CREDITS	NEW BALANCE
109.87	107.81	1.62	122.05	50.00	183.54

DATE PAYMENT DUE 10/29/--   TO AVOID ADDITIONAL FINANCE CHARGES, PAY NEW BALANCE BEFORE DATE PAYMENT DUE.

FINANCE CHARGE RATES
PERIODIC RATE PER MONTH 1.5%
ANNUAL PERCENTAGE RATE 18%

Previous balance is the money that was still owed at the end of last month. $109.87

The **average daily balance** is the same as the principal. It is the average amount of money owed over the entire month. $107.81

The **finance charge** is the same as the interest. The rate is usually listed as the percent per month. Usually the annual percentage rate is also listed on the bill. Often it is on the back of the statement. $1.62

Purchases identifies the additional purchases charged during the month. $122.05

Payments and credits identifies how much was paid toward the bill the previous month.

$50.00

The **new balance** = the (previous balance + the finance charge + the purchases) − (payments and credits).

$183.54

Date payment due is the day the company must receive the payment.

10/29

The **periodic rate per month** is the annual interest rate divided by 12.

1.5%

The **annual percentage rate** (APR) is the interest rate for the year.

18%

Look at the monthly charge account statement below, then write the answers to the questions in the blank provided.

1.

PREVIOUS BALANCE	AVERAGE DAILY BALANCE	FINANCE CHARGES	PURCHASES	PAYMENTS & CREDITS	NEW BALANCE
$1,298.34	$987.33	$19.75	$54.77	$400.00	$972.86

DATE PAYMENT DUE	3/6/-		TO AVOID ADDITIONAL FINANCE CHARGES, PAY NEW BALANCE BEFORE DATE PAYMENT DUE.
FINANCE CHARGE RATE			
ANNUAL PERCENTAGE RATE 24%			

(1) What is the annual percentage rate?

_____

(2) What is the average daily balance?

_____

(3) How much money was spent on additional purchases?

_____

(4) What is the new balance?

_____

(5) The annual percentage rate is 24%. What is the periodic rate?

_____

(6) What is the finance charge?

_____

LIFE SKILL

2.

PREVIOUS BALANCE	AVERAGE DAILY BALANCE	FINANCE CHARGES	PURCHASES	PAYMENTS & CREDITS	NEW BALANCE
$125.90	$76.50	$1.38	$37.45	$125.90	$38.83

DATE PAYMENT DUE	3/6/--		TO AVOID ADDITIONAL FINANCE CHARGES, PAY NEW BALANCE BEFORE DATE PAYMENT DUE.
FINANCE CHARGE RATE			
MONTHLY PERIODIC RATE IS 1.8%			

(1) The periodic rate is 1.8%. What is the annual percentage rate?

_____

(2) What is the average daily balance?

_____

(3) How much money was spent on additional purchases?

_____

(4) What is the new balance?

_____

(5) What is the periodic rate per month?

_____

(6) What is the finance charge?

_____

Finding the Monthly Finance Charge

It is usually a good idea to check the arithmetic on a monthly credit statement. To find the finance charge for one month, use the periodic rate on the average daily balance. Since these are decimal amounts, solve using the decimal method.

Example

Find the finance charge for the average daily balance. Then find the new balance.

PREVIOUS BALANCE	AVERAGE DAILY BALANCE	MONTHLY PERIODIC RATE	FINANCE CHARGE	NEW BALANCE
$65.00	$48.00	1.5%		

Step 1 Draw a grid.

? finance charge	1.5 Periodic rate per month
48 Average daily balance	100 100

Step 2 Find the finance charge. Multiply the diagonals

$$48 \times 1.5 = 72$$

Divide by the number that is left

$$72 \div 100 = .72$$

The finance charge is $0.72

Step 3 Add the finance charge to the previous balance to find the new balance.

$$
\begin{array}{r}
\$ 6 5 . 0 0 \\
+ \quad . 7 2 \\
\hline
\$ 6 5 . 7 2
\end{array}
$$

The new balance is $65.72.

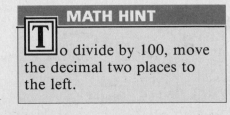

MATH HINT

To divide by 100, move the decimal two places to the left.

193

(1) Find the finance charge for the average daily balance; (2) then add it to the previous balance to find the new balance.

1.

PREVIOUS BALANCE	AVERAGE DAILY BALANCE	MONTHLY PERIODIC RATE	FINANCE CHARGE	NEW BALANCE
$431.50	$80.00	1.5%		

2.

PREVIOUS BALANCE	AVERAGE DAILY BALANCE	MONTHLY PERIODIC RATE	FINANCE CHARGE	NEW BALANCE
$75.00	$135.00	1.2%		

3.

PREVIOUS BALANCE	AVERAGE DAILY BALANCE	MONTHLY PERIODIC RATE	FINANCE CHARGE	NEW BALANCE
$1,077.32	$220.00	1.9%		

4.

PREVIOUS BALANCE	AVERAGE DAILY BALANCE	MONTHLY PERIODIC RATE	FINANCE CHARGE	NEW BALANCE
$33.45	$566.00	2%		

5.

PREVIOUS BALANCE	AVERAGE DAILY BALANCE	MONTHLY PERIODIC RATE	FINANCE CHARGE	NEW BALANCE
$185.02	$1,200.00	1.5%		

6.

PREVIOUS BALANCE	AVERAGE DAILY BALANCE	MONTHLY PERIODIC RATE	FINANCE CHARGE	NEW BALANCE
$113.22	$3,450.00	2.2%		

Problem Solving—Making a Plan

In the previous units, you have used the steps below to solve word problems. Step 2 asks you to make a plan. Remember when designing your plan to solve a percent problem, you may have to perform two or three separate calculations to arrive at the correct answer.

Step 1   Read the problem and underline the key words. These words will usually relate to some mathematics reasoning computation.

Step 2   Make a plan to solve the problem. Ask yourself, Should I add, subtract, multiply, divide, round, or compare? You may have to do **more than one of these operations** for the same problem.

Step 3   Find the solution. Use your math knowledge to find your answer.

Step 4   Check the answer. Ask yourself, Is the answer reasonable? Did you find what you were asked for?

Example

Beatriz wants to buy a stereo system for $400. The music store offers the following deal: 10% down and $30.50 a month for twelve months. What is the total cost of the stereo system?

Step 1   Determine the total cost of the stereo system. The key words are **10% down, $30.50 a month,** and **total cost.**

Step 2   The key words indicate that this is a percent problem. However, the word **total** also indicates you will have to add. The questions you must ask yourself are: What will be added? What should be done first?
   a. First, find the amount of the down payment.
   b. Then multiply the monthly payments by 12.
   c. Now add the down payment to the monthly payments. This will give you the total cost of the stereo system.

Step 3   Solve the problem.
   a. $10\% \times \$400 = \$40$ down payment
   b. $\$30.50 \times 12 = \$366$ total monthly payments
   c. $\$40 + \$366 = \$406$ total cost of the stereo system

Step 4 Check the answer. One way to make sure the answer makes sense is to recall that percent problems are used to relate different numbers to each other in terms of 100. You could use a grid to help you check part of your answer. Remember to multiply the diagonals and divide by the number that is left when using a grid.

a.

?	10
part	percent
400	100
whole	100

$$\frac{10 \times 400}{100} = \$40$$

You still have to do two more calculation checks. However, you know now that the \$40 is the correct down payment.

b. How would you check your answer for the monthly payments? You could use a calculator to recheck your figures, or you could divide. Division is one way to check a multiplication answer.

$$
\begin{array}{r}
\$30.50 \\
12)\overline{\$366.00} \\
\underline{36} \\
6\,0 \\
\underline{6\,0} \\
0
\end{array}
$$

Divide the total payments by the number of months. The answer should be the original monthly payment—and it is.

c. How would you check your answer for the total cost? Again, you could use a calculator to recheck your figures, or you could subtract. Subtraction is one way to check an addition answer.

$$
\begin{array}{r}
\$\,4\,0\,6 \\
-3\,6\,6 \\
\hline
\$4\,0
\end{array}
\quad \text{or} \quad
\begin{array}{r}
\$\,4\,0\,6 \\
-4\,0 \\
\hline
\$\,3\,6\,6
\end{array}
$$

Subtracting one of the smaller numbers from the largest should give you the third number—and it does.

Solve the following problems. Use a grid whenever possible and recheck your answers. Show all your work.

1. The Boyle family put 8% down on $588 worth of camping gear they bought. They are paying $46.50 a month for one year. (1) What was the down payment? (2) What are the total monthly payments? (3) What is the total cost of the camping gear?

(1) _____

(2) _____

(3) _____

2. Leroy has a savings account. He deposits $15 a month into his account. (1) How much has he deposited at the end of the year? (2) If the bank has an annual interest rate of 6%, how much interest will Leroy's account earn at the end of the year? (3) What will be the total in his savings account at the end of the year?

(1) _____

(2) _____

(3) _____

3. Catherine took out a loan for $8,000 from a bank for three years. The rate for the loan is 9%. (1) What is the interest on this loan at the end of three years? (2) What is the principal of this loan? (3) What is the rate? (4) What is the time?

(1) _____

(2) _____

(3) _____

(4) _____

Problem Solving

Find the amount of interest on the following:

1. $982 in a three-month savings certificate at 6% _____

2. A $280 savings account for two years at 5% _____

3. A $1,200 savings certificate for 3 months at 6.25% _____

4. A car loan for $3,472 at 9.9% for 30 months _____

5. $755 in a 5% savings account for four years _____

Read the following information and answer the questions below.

A living room set is selling for $2,500. It is advertised as having free financing with a 10% down payment. Or you can pay $225 a month on an installment plan for a year with no money down.

6. If you want free financing, how much money must be paid as a down payment?

       _____

7. What is the total cost of the living room set on the installment plan?

       _____

8. What is the total cost of a loan of $2,500 at 16% a year?

       _____

9. What is the difference in the cost of the installment plan and the loan?

       _____

10. What method of payment would you choose and why?

_____

_____

Ratios and Proportions

Solve the following proportion problems.

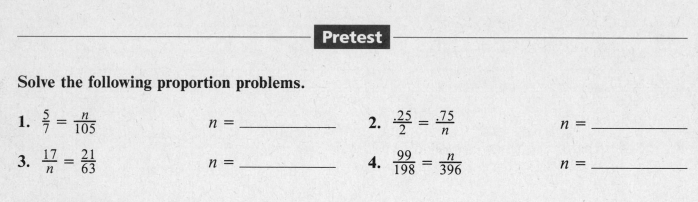

1. $\frac{5}{7} = \frac{n}{105}$ $n =$ _____   2. $\frac{.25}{2} = \frac{.75}{n}$ $n =$ _____

3. $\frac{17}{n} = \frac{21}{63}$ $n =$ _____   4. $\frac{99}{198} = \frac{n}{396}$ $n =$ _____

Problem Solving

Circle the correct answer to the following problems.

5. Nicole Price is a hairdresser. She takes care of 9 customers in 5 hours. What is the ratio of customers to hours?

    (1) $\frac{5}{9}$ **(2)** $\frac{9}{5}$

    (3) $1\frac{4}{5}$ **(4)** 5:9

6. Nicole earns \$0.75 in tips for every \$5 a customer spends in the shop. Her customers have spent \$325. What is the proportion that will tell what she earned in tips?

    (1) $\frac{\$5.00}{\$.75} = \frac{\$325}{n}$ **(2)** $\frac{\$5.00}{n} = \frac{\$.75}{\$325}$

    (3) $\frac{1}{10} = \frac{n}{\$325}$ **(4)** $\frac{10}{1} = \frac{n}{\$325}$

7. Three out of every 7 customers will have a haircut and a shampoo. If 91 people have a shampoo, how many will get their hair cut?

    (1) 13   **(2)** 21

    (3) 60   **(4)** 39

8. Nicole earns \$2 in commissions for every \$5 of business. If she brings in \$325, how much money will she get in commissions?

    (1) \$130   **(2)** \$65

    (3) \$13   **(4)** \$650

Ratios

A **ratio** is a comparison of two numbers. Study the three examples of ways to write a ratio.

MATH HINT

Ratios must always have two numbers. Never write a ratio as a mixed number.

Examples

A. Use **to, per, for, at,** or **in.**

    5 policemen per 5,000 people
    5 policemen for 5,000 people

B. Use the ratio symbol (:).

    5 policemen : 5,000 people

Notice the first term in the problem is written first when the **:** is used.

C. Use the fraction method.

$$\frac{5 \text{ policemen}}{5,000 \text{ people}}$$

Notice the first term in the problem is written on top when the fraction method is used.

When a ratio is written as a fraction, it can be reduced.

$$\frac{5 \text{ policemen}}{5,000 \text{ people}} \text{ reduces to } \frac{1 \text{ policeman}}{1,000 \text{ people}}$$

Practice

Solve the following problems. The first two are done for you.

1. Write the ratio 24 chairs to 6 tables using the three different methods.

    24 chairs to 6 tables

    24 chairs : 6 tables

$$\frac{24 \text{ chairs}}{6 \text{ tables}} \text{ or } \frac{4 \text{ chairs}}{1 \text{ table}}$$

2. Write the ratio 25 students per teacher using the : and fraction methods.

The words **a, an, each,** or **per** used before a person or thing mean **one** person or thing.

    25 students : 1 teacher

$$\frac{25 \text{ students}}{1 \text{ teacher}}$$

Use the "ratio symbol (:)" to write the following ratios.

3. two TVs to one household

   _____

4. 8 ounces of meat for each person

   _____

5. one accident to 10,000,000 miles flown

   _____

6. 24 boxes to a crate

   _____

7. one inch on a map to 25 miles

   _____

8. 20 parts alcohol to 80 parts gasoline

   _____

9. 25 miles per gallon

   _____

10. one post for every 3 feet of fence

    _____

Use the "fraction method" to write the following ratios.

11. 2 quarts of cleaner to 5 quarts of water

    _____

12. 125 people to 275 seats on a plane

    _____

13. 3 yards of cloth for each dress

    _____

14. 5 inches on a map for 65 miles

    _____

15. 24 cases of nails to 1284 pounds

    _____

16. driving 336 miles in 6 hours

    _____

Identifying Ratios

Ratios are a way to relate numbers to one another.

Ratios relate part of an amount to the whole amount.

or

Ratios relate part of an amount to another part of an amount.

--- **Example** ---

Look at the group of figures that follow. There are 12 squares, 4 circles, and 8 triangles—a total of 24 figures.

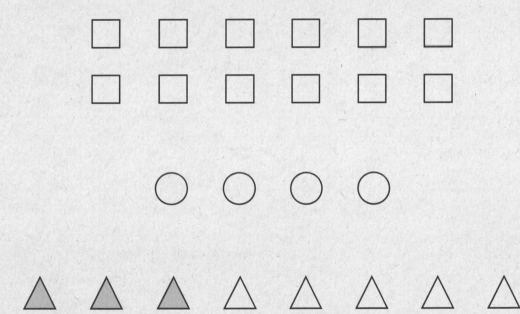

The word **figure** means the **entire group of objects.** That is the whole.

What is the ratio of triangles to all the figures?

$\frac{8}{24}$ or $\frac{1}{3}$ The ratio is the part to the whole.
There are 8 triangles to the total number of 24 figures.

What is the ratio of shaded triangles to unshaded triangles?

$\frac{3}{5}$ The ratio is part to the part.
There are 3 shaded triangles to 5 unshaded triangles.

Use the figures in the example to answer the questions that follow.
Reduce all your answers.

1. What is the ratio of squares to all the figures?

   _____

2. What is the ratio of circles to all the figures?

   _____

3. What is the ratio of shaded squares to all the squares?

   _____

4. What is the ratio of shaded circles to all the circles?

   _____

5. What is the ratio of squares to circles?

   _____

6. What is the ratio of shaded figures to unshaded figures?

   _____

7. What is the ratio of triangles to circles?

   _____

8. What is the ratio of circles to squares?

   _____

9. What is the ratio of squares to triangles?

   _____

10. What is the ratio of triangles to squares?

    _____

11. What is the ratio of circles to triangles?

12. What is the ratio of shaded squares to unshaded squares?

    _____

13. What is the ratio of shaded triangles to unshaded triangles?

    _____

14. What is the ratio of shaded circles to unshaded circles?

    _____

Identifying Ratios Using Tables

Sometimes it is helpful to organize information about ratios into a table. The table lets you see the parts and the whole, or total.

Example

Study the table that follows and determine the number of lines and columns. Read the information the columns and lines give you.

	Shaded	Unshaded	Total
Squares	6	6	12
Circles	1	3	4
Triangles	3	5	8
TOTAL	10	14	24

There are four columns in this table.

Column 1 lists the types of figures and the total.
    Line 1 states squares.
    Line 2 states circles.
    Line 3 states triangles.
    Line 4 states totals.

Column 2 tells you the number of shaded figures.

Column 3 tells you the number of unshaded figures.

Column 4 gives the totals.

What is the ratio of unshaded circles to all the figures?

Look at the table. Find the line that states circles. Find the column that tells you the number of unshaded figures and the column that tells you the number of total figures.

$$\frac{3 \text{ unshaded circles}}{24 \text{ total figures}} = \frac{1}{8}$$

The ratio of unshaded circles to all the figures is $\frac{1}{8}$, or 1:8.

Look at the figures in the box that follows. Using the information given, fill out the chart.

	Shaded	Unshaded	Total
Squares			
Circles			
Triangles			
Diamonds			
TOTAL			

Problem Solving

Now write the ratios for the following problems. Reduce all your answers.

1. What is the ratio of squares to all the figures?

2. What is the ratio of circles to all the figures?

3. What is the ratio of shaded diamonds to all the diamonds?

_____

4. What is the ratio of shaded circles to all the circles?

_____

5. What is the ratio of squares to diamonds?

_____

6. What is the ratio of shaded figures to unshaded figures?

_____

7. What is the ratio of triangles to circles?

_____

8. What is the ratio of diamonds to circles?

_____

9. What is the ratio of squares to triangles?

_____

10. What is the ratio of triangles to squares?

_____

11. What is the ratio of circles to triangles?

_____

12. What is the ratio of shaded squares to unshaded squares?

_____

Writing Proportions

A proportion is a statement that two ratios are equal. Proportions can be written in two ways:

6:8::3:4   This proportion
             is read, 6 is to
    or   8 as 3 is to 4.

$\frac{6}{8} = \frac{3}{4}$

Notice the proportion symbol (::).

Examples

A. Write the proportion
    1 is to 2 as 8 is to 16.

    $\frac{1}{2} = \frac{8}{16}$ or 1:2 :: 8:16

B. Write the proportion
    2 is to 3 as 34 is to 51.

    $\frac{2}{3} = \frac{34}{51}$ or 2:3 :: 34:51

C. Write the proportion
    108 is to 12 as 27 is to 3.

    $\frac{108}{12} = \frac{27}{3}$ or 108:12 :: 27:3

Using an equal sign is the more common way of writing proportions.

Practice

Write the following proportions using both methods.

1. 12 is to 18 as 44 is to 66

    _____   _____

2. 27 is to 54 as 45 is to 90

    _____   _____

3. 9 is to 81 as 15 is to 135

    _____   _____

4. 10 is to 100 as 100 is to 1,000

    _____   _____

5. 150 is to 300 as 75 is to 150

    _____   _____

6. 99 is to 1 as 990 is to 10

    _____   _____

7. 348 is to 58 as 162 is to 27

    _____   _____

8. 17 is to 51 as 1 is to 3

    _____   _____

Solving Proportions

The following are the **cross-products** in a proportion. Cross-products are equal.

$\frac{6}{8} = \frac{3}{4}$ $6 \times 4 = 24$

$\frac{2}{3} = \frac{8}{12}$ $3 \times 8 = 24$
$2 \times 12 = 24$

$\frac{6}{8} = \frac{3}{4}$ $8 \times 3 = 24$

Check these proportions to see if their cross-products are equal.

$\frac{15}{45} = \frac{17}{51}$ $15 \times 51 = 765$
$45 \times 17 = 765$

$\frac{2}{9} = \frac{18}{81}$ $2 \times 81 = 162$
$9 \times 18 = 162$

To solve a proportion means to find a missing number in the proportion. To find the missing number, use the following steps:

Step 1 Let the letter n stand for the missing number.

Step 2 Find the cross-product of the numbers you know.

Step 3 Divide the cross-product by the number that is left.

> **MATH HINT**
>
> The cross-products in a proportion are equal.
>
> $\frac{2}{3} = \frac{8}{12}$ $3 \times 8 = 24$
> $2 \times 12 = 24$

Examples

A. Solve the proportion. $\frac{3}{4} = \frac{?}{16}$

Step 1 Let the letter n stand for the missing number.

$$\frac{3}{4} = \frac{n}{16}$$

Step 2 Find the cross-product of the numbers you know.

$$3 \times 16 = 48$$

Step 3 Divide the cross-product by the number that is left.

$$\frac{48}{4} = 12 \qquad n = 12$$

B. One-half cup flour is used for a recipe that makes four servings. How much flour is needed for 12 servings?

Step 1 $\quad \dfrac{\frac{1}{2}}{4} = \dfrac{n}{12}$

Step 2 $\quad \dfrac{1}{2} \times 12 = 6$

Step 3 $\quad \dfrac{6}{4} = 1\dfrac{2}{4} = 1\dfrac{1}{2}$ cups

C. 1.5 pounds of asparagus sell for \$3.75. What is the price of one pound of asparagus?

Step 1 $\quad \dfrac{1.5}{3.75} = \dfrac{1}{n}$

Step 2 $\quad 3.75 \times 1 = 3.75$

Step 3
$$
\begin{array}{r}
2.50 \\
1.5\overline{)3.75} \\
3\,0 \\
\overline{75} \\
75 \\
\overline{0}
\end{array}
$$

$n = \$2.50$ a pound

> **MATH HINT**
>
> **N**otice that order is important. The second ratio in a proportion must be in the same order as the first ratio.
>
> $\dfrac{\text{eggs}}{\text{dollars}} = \dfrac{\text{eggs}}{\text{dollars}}, \dfrac{\text{sugar}}{\text{recipe}} = \dfrac{\text{sugar}}{\text{recipe}}$

Practice

Check these proportions to see if their cross-products are equal.

1. $\dfrac{8}{24} = \dfrac{9}{27}$ $8 \times 27 = $ _____

$9 \times 24 = $ _____

2. $\dfrac{3}{15} = \dfrac{7}{35}$ $3 \times 35 = $ _____

$7 \times 15 = $ _____

Find each missing number.

3. $\dfrac{5}{8} = \dfrac{n}{88}$ $n = $ _____

4. $\dfrac{12}{96} = \dfrac{n}{64}$ $n = $ _____

5. $\dfrac{51}{n} = \dfrac{17}{1}$ $n = $ _____

6. $\dfrac{11}{n} = \dfrac{32}{96}$ $n = $ _____

7. $\dfrac{n}{48} = \dfrac{7}{8}$ $n = $ _____

8. $\dfrac{n}{144} = \dfrac{7}{84}$ $n = $ _____

9. $\dfrac{n}{8} = \dfrac{\frac{3}{8}}{4}$ $n = $ _____

10. $\dfrac{\frac{1}{8}}{88} = \dfrac{11}{n}$ $n = $ _____

11. $\dfrac{11}{3} = \dfrac{n}{\frac{1}{6}}$ $n = $ _____

12. $\dfrac{\frac{3}{5}}{5} = \dfrac{n}{25}$ $n = $ _____

13. $\dfrac{.8}{1.6} = \dfrac{.1}{n}$ $n = $ _____

14. $\dfrac{1.5}{3} = \dfrac{n}{5}$ $n = $ _____

LIFE SKILL

Reading a Map

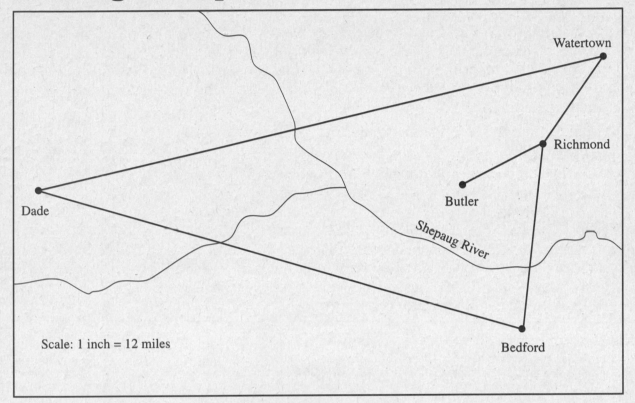

Scale: 1 inch = 12 miles

The distance from Dade to Watertown is 6 inches on the map. How far apart are the towns in miles?

$$\frac{1 \text{ in.}}{12 \text{ mi}} = \frac{6 \text{ in.}}{n}$$

$$n = \frac{12 \text{ mi} \times 6}{1}$$

$$= 72 \text{ mi}$$

1. It's 12 miles from Butler to Richmond. How far apart are they on the map?

   _____

2. Ted Sims was making deliveries. His first stop was 9 miles from Butler. How many inches would that be on the map?

   _____

3. On his way from Richmond to Bedford, he stopped. On the map, it looked like $\frac{1}{2}$ inch. How many miles had he gone?

   _____

4. A tire went flat just as he crossed the Shepaug River. On the map, the river crossed the road $\frac{5}{8}$ inch away from Bedford. How far away from the town was he?

   _____

Problem Solving—Proportions

Use the following steps to help you solve proportion word problems. Remember, in proportion problems you are working with two numbers related to one another.

Step 1 Read the problem and underline the key words for the two things being compared. Write the two things being compared one over the other. Then, write the proportion in the same order. Remember the variable n is used for the missing number.

Step 2 Make a plan to solve the problem. You should solve using cross-products.

Step 3 Find the solution. Use your math knowledge to find your answer.

Step 4 Label your answer.

Also, when using the above steps, make sure that the two things being compared in a proportion problem remain **in the same order**. Here are some key words that indicate proportion problems are two numbers related to one another: **rate, per, in, at, to.**

Example

Solve for the proportion in the following word problem.
Hank can pack 26 egg cartons per cardboard box. How many cardboard boxes will he need to hold 3,900 egg cartons?

Step 1 Read the problem and identify the key word, which is **per.** The two things being compared one over the other are egg cartons to cardboard boxes. If you write the proportion in the same order, it should look like this:

$$\frac{\text{egg cartons}}{\text{cardboard boxes}} \qquad \frac{26}{1}$$

Step 2 You should solve using cross-products.

$$\frac{26}{1} = \frac{3,900}{n}$$

Step 3 Find the solution. Use your math knowledge to find your answer.

$$\frac{1 \times 3,900}{26} = \frac{3,900}{26} = 150$$

Step 4   Label your answer.

$n = 150$ cardboard boxes

Hank will need 150 cardboard boxes in order to pack 26 egg cartons.

Practice

Solve the following problems.

1. A town wants to keep its ratio of 7 small parks to 5,000 people. If the town grows to 15,000 people, how many parks will it need?

   21

2. Rudy must make 23 radio parts in 40 minutes at work. How many will he make by the end of an 8-hour day? (Hint: Express the minutes as hours.)

   276

3. Roberto used 33 gallons of gasoline on his 1,000-mile trip through the Rockies. How much gasoline will he need for a 250-mile trip?

   8.25

4. If a plane flew 932 miles in 2 hours, how far will it fly in 7 hours?

   3262

Changing a Recipe

Read the following recipe.

> ### Spaghetti / 4 servings
> Brown in a heavy pot:
> $\frac{1}{2}$ pound sausage $\frac{1}{3}$ pound hamburger
> 4 tablespoons chopped onion
> Add:
> 3 ounces tomato paste   1 teaspoon sugar
> 8 ounces tomato sauce $1\frac{1}{2}$ teaspoons salt
> 16 ounces canned whole $\frac{1}{2}$ teaspoon basil
>   tomatoes   2 teaspoons oregano
> Simmer 2 hours. Serve over 1 pound of boiled spaghetti noodles.

Solve.

If the Rowes need only 3 servings, how much of each ingredient should they use?

1. Sausage _____ 3/8 _____

2. Hamburger _____ 1/4 _____

3. Onion _____ 3 _____

4. Tomato paste _____

5. Tomato sauce _____ 2.25 _____

6. Whole tomatoes _____ 12 _____

7. Sugar _____ 3/4 _____

8. Salt _____

9. Basil _____ 3/8 _____

10. Oregano _____

The Gilmores want to cook enough for 10 servings. How much of each ingredient do they need?

11. Sausage _____

12. Hamburger _____

13. Onion _____

14. Tomato paste _____

15. Tomato sauce _____

16. Whole tomatoes _____

17. Sugar _____

18. Salt _____

19. Basil _____

20. Oregano _____

LIFE SKILL

Finding Prices by Proportion

To solve proportion problems that deal with prices, follow these steps:

Step 1 Write a proportion.

Step 2 Solve for *n*.

Step 3 Round up to the next penny if there is a remainder.

A. 2 bags of carrots cost $.98.
What is the cost of 1 bag of carrots?

$$\frac{2}{\$.98} = \frac{1}{n} \qquad \frac{\$.98 \times 1}{2} = \$.49$$

$$n = \$.49$$

B. 4 cans of beets cost $1.08.
What is the cost of 3 cans of beets?

$$\frac{4}{\$1.08} = \frac{3}{n} \qquad \frac{\$1.08 \times 3}{4} = \$.80$$

$$n = \$.81$$

Find the cost of the following items using these prices.

> 5 pounds of potatoes for $4.00
> 2 heads of lettuce for $1.44
> 3 cans of tomatoes for $2.07
> 6 ounces of bean sprouts for $.66

> 3 bags of candy for $5.55
> 2 cans of soup for $1.64
> 2 pounds of hamburger for $3.28

1. 10 pounds of potatoes _____ $8

2. 3 heads of lettuce _____ 3.66

3. 9 cans of tomatoes _____

4. 18 ounces of bean sprouts _____

5. 4 cans of soup _____

6. 2 pounds of potatoes _____

7. 3 pounds of hamburger _____

8. 16 cans of soup _____

Posttest

Solve the following proportion problems.

1. $\frac{7}{98} = \frac{n}{70}$ $\qquad$ $n =$ _____

2. $\frac{n}{.5} = \frac{15}{75}$ $\qquad$ $n =$ _____

3. $\frac{2}{5} = \frac{50}{n}$ $\qquad$ $n =$ _____

4. $\frac{53}{n} = \frac{33}{66}$ $\qquad$ $n =$ _____

5. $\frac{\frac{1}{2}}{\frac{1}{4}} = \frac{n}{\frac{3}{4}}$ $\qquad$ $n =$ _____

6. $\frac{\frac{3}{5}}{7} = \frac{n}{35}$ $\qquad$ $n =$ _____

Problem Solving

Circle the correct answer to the following problems.

7. Cal Adams works in a warehouse. He packed 63 cartons in the last 3 crates on hand. He has 1,386 more cartons to pack. How many more crates does he need?

   (1) 66   **(2)** 45   **(3)** 4.5   **(4)** 225

8. In 4 hours he will pack 18 crates. How many crates will he pack in 6 hours?

   (1) 108   **(2)** 24   **(3)** $4\frac{1}{2}$ **(4)** 27

9. Seven crates fit on a platform. The crates on each platform weigh a total of 420 pounds. How much will 98 crates weigh?

   (1) 60 pounds   **(2)** 5,880 pounds
   (3) 686 pounds   **(4)** 40,160 pounds

10. Hector works 3 hours overtime for every 40 hours on the job. How many hours of overtime will he work at the end of 240 hours on the job?

    (1) 120   **(2)** 6   **(3)** 18   **(4)** $16\frac{2}{3}$

Practice for Mastering Percents and Proportions

Change the percent, decimal, or fraction to the other forms.

Percent	Decimal	Fraction
_____15%_____	1. ____.15____	2. _____
3. ___50%___	____.50____	4. ___$\frac{1}{2}$___

Find the percent, part, or whole in the following problems.

5. What is 17% of 98? _____   6. 26 is $66\frac{2}{3}$% of what? _____

7. 57% of 1,000 equals what? _____   8. $\frac{1}{2}$ is what percent of $\frac{3}{4}$? _____

Problem Solving

Solve the following problems.

9. Norma made 150 phone calls for her job as a salesperson. She discovered that 5% of the people she called refused to speak with her. How many would talk with her? _____

10. Early on election night 1,595 votes had been counted. That was 55% of the total. What was the total number of votes? _____

11. In District One, 50% of the total vote was counted. On that day, 1,978 people in District One voted. How many of the ballots have been counted? _____

12. Tom's teenage son Jim pays 85% of the cost of his clothes, transportation, and entertainment. That amounts to all of Jim's weekly pay, $30.60, from his part-time job. What is the amount of Jim's weekly expenses? _____

13. The cost of a bus ride is $1.25. The fare is being raised a nickel. What percent of the original price is the raise? _____

14. Miguel is setting up a small business. He will be purchasing office equipment for himself and his staff.

(1) The desks he wants are $340 each marked down from $500. What is the percent of discount?

_____

(2) The chairs with adjustable back rests are $250 each. The ones without adjustable back rests are $200 each. What is the percent of increase for the chairs with adjustable back rests?

_____

(3) The lamps, desk accessories and supplies are $380. That brings the total amount of the purchase to $2,150. On the installment plan he can put 20% down and pay $150 a month for 12 months. What is the total cost of the installment plan?

_____

(4) He can get a loan for $2,150 for one year at 14% interest. What is the cost of the loan?

_____

(5) Which would be less expensive, the loan or the installment plan?

_____

(6) The start up costs of his business are $25,000. Thirty percent of that is spent on the down payment for his lease, 50% on equipment, and the rest on advertising. How much does he spend on advertising?

_____

(7) His income is expected to be $12,000 a month. Twenty percent of that will come from direct sales, 40% from catalog sales, and the rest equally divided between telemarketing and cold calls. How much money does he expect to get from telemarketing?

_____

(8) The ratio of successful sales to numbers of phone calls made is $\frac{1}{500}$. How many calls must his staff make to get 15 sales?

_____

15. The Magelenos wanted to celebrate their wedding anniversary. They spoke to different banquet halls to find out what a dinner for 150 people would cost. One place said they would charge $30 a plate for a total of $4,500. The installment plan they had required 25% down and $350 a month for a year.

(1) What was the amount of the down payment? _____

(2) What was the total cost of the installment plan? _____

(3) If they took out a loan for $4,500 at 18% for one year, what would be the interest on the loan? _____

(4) What is the total cost of the loan? _____

(5) Which is more expensive, the loan or the installment plan? _____

16. The first weekend Carlos made $60 in tips working at the golf club. The second weekend he made $90. What is the percent of increase from the first weekend to the second weekend? _____

17. The Lopez family was looking at its budget. They found that 56% of the family's money was spent on the mortgage payment and utilities, 10% was put into savings, and 14% was spent on food. The rest was equally spent on transportation and clothes.

(1) What percent was spent on clothes? _____

(2) If the family's annual income is $48,000, how much money is spent on transportation? _____

Pretest/Unit 1/pages 1-2

1. 10.77
2. 195.64
3. 42.98
4. 948.69
5. 12.89
6. 9,024.08
7. 8.01
8. 0.06
9. 222,592
10. 2.3475
11. 1,283.72
12. 35.62
13. 600
14. 20
15. 650
16. 0.04, $\frac{1}{4}$, 0.4
17. 0.08, $\frac{1}{3}$, $\frac{7}{8}$
18. $5\frac{4}{5}$
19. $7\frac{2}{7}$
20. $3\frac{23}{24}$
21. $11\frac{3}{4}$
22. $3\frac{2}{9}$
23. $10\frac{3}{7}$
24. $9\frac{7}{8}$
25. $3\frac{1}{10}$
26. $25\frac{1}{4}$
27. 45
28. 15
29. 30
30. $33\frac{1}{3}\%$
31. $12\frac{1}{2}\%$

Lesson 1/pages 3-4

1. 8.63
2. 459.191
3. 1.68
4. 1,023.05
5. 11.16
6. 567.03
7. 787.23
8. 16.543
9. 0.09
10. 1,033.88
11. 5 minutes
12. $17.74
13. $44

Lesson 2/pages 5-6

1. 13.59
2. 34.506
3. 0.094
4. 0.55
5. 1.177
6. 10.5041
7. 976.8
8. 82.65
9. 2.2533
10. 2,306.5082
11. 161.633
12. 0.9610755
13. $41.55
14. $31.20
15. $676.40

Lesson 3/pages 7-8

1. 600
2. 28
3. 700
4. 8.1
5. 0.45
6. 7,100
7. 40
8. 50
9. 70
10. 20 groups
11. 100 spools
12. $3,400

Lesson 4/pages 9-11

1. $\frac{1}{12}$
2. $\frac{1}{20}$
3. $8\frac{1}{4}$
4. $10\frac{2}{3}$
5. $11\frac{1}{2}$
6. $17\frac{1}{3}$
7. $\frac{15}{25}$
8. $\frac{49}{56}$
9. $\frac{9}{81}$
10. $\frac{60}{72}$
11. $\frac{45}{55}$
12. $\frac{6}{39}$
13. 9
14. 20
15. 16

16. 60
17. 18
18. 60
19. $8\frac{3}{2}$
20. $5\frac{3}{3}$
21. $8\frac{12}{12}$
22. $20\frac{15}{8}$
23. $46\frac{11}{8}$
24. $18\frac{3}{2}$
25. $4\frac{2}{5}$
26. 1
27. $7\frac{2}{4} = 7\frac{1}{2}$
28. $5\frac{3}{9} = 5\frac{1}{3}$
29. 9
30. 1
31. $\frac{27}{5}$
32. $\frac{37}{4}$
33. $\frac{100}{3}$
34. $\frac{67}{4}$
35. $\frac{15}{2}$
36. $\frac{25}{3}$

Lesson 5/pages 12-13

1. $\frac{1}{6}$
2. $1\frac{1}{3}$
3. $\frac{4}{39}$
4. $3\frac{1}{8}$
5. $12\frac{1}{2}$
6. $\frac{7}{40}$
7. 18
8. $1\frac{3}{5}$
9. 50 feet
10. $21\frac{1}{3}$ yards
11. $22\frac{1}{2}$ feet
12. $45\frac{1}{2}$ inches

Lesson 6/pages 14-15

1. 1
2. $1\frac{1}{2}$
3. $1\frac{2}{9}$
4. 44
5. 20
6. $\frac{1}{8}$
7. $\frac{3}{4}$
8. $\frac{1}{4}$
9. $2\frac{3}{4}$
10. $\frac{2}{3}$
11. 141
12. 2
13. 160 flowers
14. 40 markers
15. $1\frac{1}{6}$ feet
16. $6\frac{1}{6}$ feet

Lesson 7/pages 16-18

1. $3\frac{3}{5}$
2. $11\frac{1}{3}$
3. $5\frac{3}{8}$
4. $12\frac{3}{10}$
5. 3
6. $17\frac{13}{35}$
7. $13\frac{7}{12}$
8. $13\frac{7}{8}$
9. $22\frac{2}{33}$
10. $28\frac{13}{33}$
11. 17
12. 13
13. $10\frac{5}{12}$
14. $20\frac{1}{72}$
15. $18\frac{26}{33}$
16. $52\frac{11}{12}$
17. $7\frac{31}{35}$
18. $9\frac{3}{10}$
19. (2)
20. (1)

Lesson 8/pages 19-20

1. 0.1
2. 5.375
3. $0.44\frac{4}{9}$
4. 0.875
5. 1.5
6. $0.66\frac{2}{3}$
7. 0.8
8. 0.4375
9. $.16\frac{2}{3}$
10. 0.75
11. $0.11\frac{1}{9}$
12. 0.3
13. 0.4
14. 11.9
15. $0.33\frac{1}{3}$
16. $0.83\frac{1}{3}$
17. $\frac{1}{8}$
18. $\frac{1}{4}$
19. $\frac{9}{40}$
20. $\frac{8}{25}$
21. $5\frac{1}{100}$
22. $\frac{1}{5}$
23. $7\frac{1}{5}$
24. $\frac{1}{1,000}$
25. $\frac{17}{20}$
26. $\frac{3}{1,000}$
27. $\frac{5}{8}$
28. $4\frac{1}{10}$
29. $\frac{3}{2,500}$
30. $6\frac{1}{8}$
31. $\frac{503}{1000}$
32. $5\frac{9}{100}$

Lesson 9/pages 21-22

1. $\frac{3}{4}$
2. 0.3
3. 1.1
4. 0.85
5. 0.4
6. 0.75
7. 0.09
8. 0.009
9. 0.49

10. $2\frac{4}{5}$
11. $\frac{1}{8}$
12. $\frac{1}{2}$
13. 0.09 $\frac{1}{5}$ 0.9
14. $\frac{3}{8}$ $\frac{5}{9}$ 0.6
15. 0.32 .23 .32
16. 0.025 .1025 .125
17. $\frac{2}{4}$ $\frac{3}{4}$ $\frac{7}{8}$
18. $\frac{1}{8}$ 0.4 $\frac{1}{2}$
19. 0.05 0.5 1.6
20. 0.02 0.2 11.5

Lesson 10/pages 23-24

1. $13\frac{1}{2}$ hours
2. $12\frac{3}{8}$ feet
3. $2\frac{5}{8}$ yards
4. $54.66
5. $17.20 (excluding tax and tip)

Posttest/Unit 1/pages 25-26

1. 9.37
2. 298.57
3. 88.66
4. 637.94
5. 44.96
6. 5,092.58
7. 1.71
8. $25\frac{17}{64}$
9. 0.08
10. 1.8312
11. 1,814.92
12. 83.694
13. 4,020
14. $\frac{1}{9}$
15. 614.28571
16. 0.05 $\frac{1}{5}$ 0.50
17. 0.09 $\frac{8}{9}$ $\frac{9}{8}$
18. $9\frac{3}{4}$
19. $5\frac{3}{5}$
20. $12\frac{1}{12}$
21. $630.95
22. $2\frac{19}{20}$ feet
23. $4,000, $3,000, $5,000

Pretest/Unit 2/pages 27-28

1. 20%
2. 80%
3. 100%

4. 0.25
5. 2.45
6. 0.01
7. $0.66\frac{2}{3}$
8. 0.77
9. 0.0008
10. $\frac{9}{10}$
11. $\frac{1}{3}$
12. $2\frac{22}{25}$
13. $\frac{17}{25}$
14. $\frac{1}{200}$
15. $\frac{1}{25}$
16. $33\frac{1}{3}\%$
17. 175%
18. 4,210%
19. 100%
20. 0.82%
21. 20%
22. $\frac{1}{4}$, 30%, .75, $\frac{4}{5}$
23. 15%, $\frac{1}{4}$, .33, $\frac{2}{5}$
24. .04, 18%, $\frac{2}{4}$, 2.1
25. $\frac{2}{3}$, 35%, $\frac{1}{3}$, 30%
26. $\frac{5}{8}$, $\frac{6}{10}$, $\frac{2}{4}$, $\frac{1}{5}$
27. 0.4, $\frac{1}{4}$, $\frac{2}{10}$, 0.04
28. 50%
29. 20%
30. 25%

Lesson 11/pages 29-32

1. (1) The container of syrup.
   (2) 100% natural means all the ingredients in the syrup are natural.
2. (1) Loan requirements.
   (2) 0% down payment means no money down.
3. (1) The box of vitamins.
   (2) 100% of the minimum daily requirements means the product provides all the minimum daily requirements established by the government.
4. 0%
5. 100%
6. 100%
7. 100%
8. 0%
9. 100%
10. 100%
11. 100%

12. 100%
13. 100%
14. 100%
15. 100%

Life Skill/page 33

1. Answers will vary depending on how many servings of each food group was selected.
   Sample answer:

Hot Farina Cereal	45%
Beans	25%
Spaghetti	25%
Rice Pudding	5%
	100%

Lesson 12/pages 34-37

1. 75%
2. 25%
3. 40%
4. 60%
5. 100%
6. 4%
7. 96%
8. 100%
9. 35%
10. 65%
11. 100%
12. 60%
13. 40%
14. 100%
15. The entire amount or the total of an amount equals 100%.

Lesson 13/pages 38-40

1. (1) $99\frac{1}{4}\%$
   (2) $\frac{3}{4}\%$
   (3) 100%
2. (1) 99.25%
   (2) 0.75%
   (3) No
   (4) Answers will vary. Sample Answer: One is measured in fractions. One is measured in decimals.
   (5) They are equal.
3. 0.5%
4. $\frac{1}{5}\%$
5. 8%
6. Subtract 0.4% from 100%
7. 9%
8. any amount representing less than $\frac{1}{100}$th of the whole

Lesson 14/pages 41-42

1.

2.

Lesson 15/page 43

1. 200%   6. 400%
2. 300%   7. 500%
3. 300%   8. 200%
4. 400%   9. 300%
5. 200%   10. 300%

Lesson 16/pages 44-45

1.
45%

2.
75%

3.
195%

4.
5%

5.
80%

6.
132%

7.
100%

8.
175%

9.
20%

10.
60%

Lesson 17/pages 46-47

1. 0.25
2. 0.75
3. 0.4
4. 0.8
5. 0.3
6. 0.9
7. 0.2
8. 100%
9. 25%
10. 20%
11. 70%
12. 90%
13. 60%
14. 10%
15. Answers may vary. Sample Answer: Decimals represent the parts of a whole in multiples of tens using the decimal point. Percents represent parts of a whole in 100ths using a percent sign.

Lesson 18/pages 48-49

1. $\frac{25}{100} = \frac{1}{4}$
2. $\frac{75}{100} = \frac{3}{4}$
3. $\frac{40}{100} = \frac{2}{5}$
4. $\frac{80}{100} = \frac{4}{5}$
5. $\frac{30}{100} = \frac{3}{10}$
6. $\frac{90}{100} = \frac{9}{10}$
7. $\frac{20}{100} = \frac{1}{5}$
8. 50%
9. 25%
10. 20%
11. 80%
12. 40%
13. 70%
14. 30%
15. Answers will vary. Sample Answer: Fractions represent parts of a whole unit divided into any number of parts represented by the numerator and denominator. Percents represent parts of a whole in terms of 100 using a percent sign.

Life Skill/page 50

1. (2)
2. (1)
3. (2)
4. (2)
5. (1)
6. (1)

Lesson 19/pages 51-53

1. $\frac{1}{2}$ dollar
2. 0.5
3. 50¢
4. 0.10 or 0.1
5. $0.50
6. 10¢
7. $\frac{1}{10}$ of a dollar
8. $0.10
9. 50%
10. 75%
11. 50%
12. 75%
13.

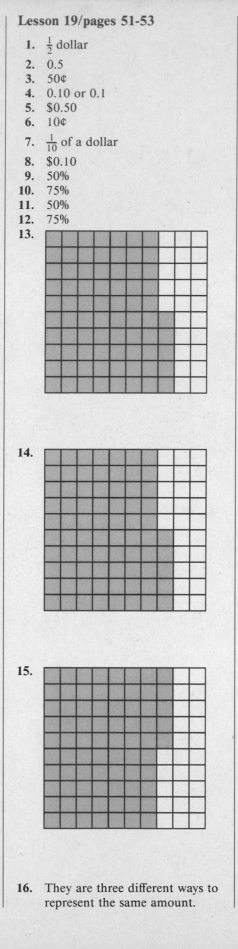

14.

15.

16. They are three different ways to represent the same amount.

17.

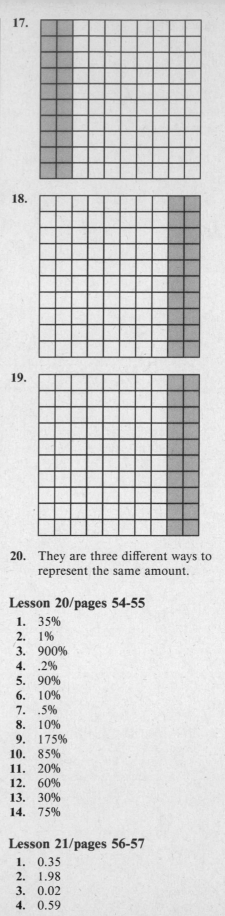

18.

19.

20. They are three different ways to represent the same amount.

Lesson 20/pages 54-55

1. 35%
2. 1%
3. 900%
4. .2%
5. 90%
6. 10%
7. .5%
8. 10%
9. 175%
10. 85%
11. 20%
12. 60%
13. 30%
14. 75%

Lesson 21/pages 56-57

1. 0.35
2. 1.98
3. 0.02
4. 0.59

5. 9.99
6. 0.007
7. 0.05
8. 1.4
9. 0.9
10. 0.75
11. 0.19
12. 0.32
13. 0.5
14. 0.6
15. 0.83

Lesson 22/pages 58-59

1. $\frac{9}{10}$
2. $\frac{13}{100}$
3. $1\frac{9}{10}$
4. $\frac{11}{50}$
5. $\frac{3}{5}$
6. $\frac{17}{20}$
7. $1\frac{1}{2}$
8. $\frac{1}{2}$
9. $\frac{1}{4}$
10. 1
11. $\frac{2}{3}$
12. $\frac{3}{8}$
13. $\frac{1}{6}$
14. $\frac{7}{8}$
15. $\frac{5}{6}$
16. $\frac{33}{200}$
17. $\frac{17}{80}$
18. $\frac{1}{12}$
19. $\frac{1}{4}$
20. $\frac{4}{5}$
21. $\frac{3}{4}$
22. $\frac{1}{5}$
23. 3
24. $\frac{3}{5}$
25. $\frac{1}{5}$

Lesson 23/pages 60-61

1. 40%
2. 25%
3. 4%

4. 290%
5. 15%
6. $33\frac{1}{3}\%$
7. $83\frac{1}{3}\%$
8. $37\frac{1}{2}\%$
9. $16\frac{2}{3}\%$
10. $62\frac{1}{2}\%$
11. $87\frac{1}{2}\%$
12. $8\frac{1}{3}\%$
13. $14\frac{2}{7}\%$
14. $71\frac{3}{7}\%$
15. $85\frac{5}{7}\%$
16. 20%
17. 75%
18. 350%
19. 80%
20. 16%

Lesson 24/pages 62-63

	Percent	Fraction	Decimal
1.	$33\frac{1}{3}\%$	$\frac{1}{3}$	$0.33\frac{1}{3}$
2.	10%	$\frac{1}{10}$	0.1
3.	$16\frac{2}{3}\%$	$\frac{1}{6}$	$0.16\frac{2}{3}$
4.	135%	$1\frac{7}{20}$	1.35
5.	62.5% or $62\frac{1}{2}\%$	$\frac{5}{8}$	0.625
6.	90%	$\frac{9}{10}$	0.90
7.	$\frac{1}{5}\%$ or .2%	$\frac{1}{500}$	0.002
8.	$1\frac{1}{2}\%$	$\frac{3}{200}$	0.015
9.	99%	$\frac{99}{100}$	0.99
10.	2,557%	$25\frac{57}{100}$	25.57
11.	20%	$\frac{1}{5}$	0.20
12.	75%	$\frac{3}{4}$	0.75
13.	$66\frac{2}{3}\%$	$\frac{2}{3}$	$0.66\frac{2}{3}$
14.	$12\frac{1}{2}\%$	$\frac{1}{8}$	$0.12\frac{1}{2}$
15.	60%	$\frac{6}{10}$ or $\frac{2}{5}$	0.60

Life Skill/pages 64-65

1. E
2. C
3. A
4. B
5. D

Lesson 25/pages 66-67

1. 0.8
2. 15%
3. 0.22
4. $\frac{3}{4}$
5. 1
6. 1
7. 0.6
8. $\frac{1}{2}$
9. 33%
10. 85%
11. 100%
12. 99%
13. 5%
14. $\frac{2}{5}$
15. 55%
16. 20%
17. 14%
18. 49%
19. 0.3
20. $\frac{1}{10}$

Lesson 26/pages 68-69

1. (1) 0.1
   (2) 0.09
   (3) 0.02
   (4) 0.1 > 0.09 > .02
2. (1) 1.00
   (2) 1.5
   (3) 1.45
   (4) 1.5 > 1.45 > 1
3. (1) 0.33
   (2) 0.035
   (3) 0.3
   (4) 0.33 > 0.3 > 0.035
4. (1) $\frac{55}{100}$
   (2) $\frac{60}{100}$
   (3) $\frac{70}{100}$
   (4) $\frac{55}{100} < \frac{60}{100} < \frac{70}{100}$
5. (1) $\frac{95}{100}$
   (2) $\frac{99}{100}$
   (3) $\frac{90}{100}$
   (4) $\frac{90}{100} < \frac{95}{100} < \frac{99}{100}$
6. (1) $\frac{1}{100}$
   (2) $\frac{100}{100}$
   (3) $\frac{25}{100}$
   (4) $\frac{1}{100} < \frac{25}{100} < \frac{100}{100}$ or
   $\frac{1}{100} < \frac{1}{4} < 1$

7. (1) 40%
   (2) 20%
   (3) 75%
   (4) 75% > 40% > 20%
8. (1) 50%
   (2) $33\frac{1}{3}$%
   (3) $37\frac{1}{2}$%
   (4) 50% > $37\frac{1}{2}$% > $33\frac{1}{3}$%
9. (1) 60%
   (2) 75%
   (3) 25%
   (4) 75% > 60% > 25%

Lesson 27/pages 70-72

1. 0%
2. 100%
3. 0%
4. 0%
5. 40%
6. 100%
7. 1%
8. 78%
9. 60%
10. 55%
11. 84%
12. 15%
13. 35%
14. 98%
15. 68%

Posttest/Unit 2/pages 73-74

1. 16%
2. 84%
3. 100%

	Fraction	Decimal	Percent
4.	$\frac{19}{20}$	0.95	95%
5.	1	1.00	100%
6.	$\frac{1}{4}$	0.25	25%
7.	$2\frac{4}{5}$	2.8	280%
8.	$\frac{3}{500}$	0.006	0.6%
9.	$\frac{2}{5}$	0.4	40%
10.	$\frac{2}{125}$	0.016	1.6%
11.	$\frac{1}{2}$	0.5	50%
12.	$\frac{4}{25}$	0.16	16%

13. 0.83, $\frac{4}{5}$, 8.4%
14. 0.5, $\frac{1}{5}$, .11%
15. 0.55, $\frac{1}{2}$, .55%
16. 0.35, $\frac{1}{3}$, 33%
17. 6.3, 63%, $\frac{6}{10}$

18. 4.55, $4\frac{1}{2}$, 4.2%
19. 20%
20. 78%
21. 0.0093
22. 100%
23. 95%
24. 89%
25. 13%
26. 23%

Pretest/Unit 3/pages 75-76

1. 440
2. 30
3. $8\frac{1}{3}$%
4. 750
5. 6,000
6. 9,000
7. $66\frac{2}{3}$%
8. 200
9. 0.08
10. 300%
11. 78%
12. $100
13. $500
14. $125
15. $104.93

Lesson 28/pages 77-78

1.

82	25
328	100

2.

50	25
200	100

3.

133	20
665	100

4.

19	2
950	100

5.

6	$33\frac{1}{3}$
18	100

6.

1.5	1.5
100	100

7.

84	350
24	100

8.

5	$\frac{1}{2}$
1,000	100

9.

22	200
11	100

10.

300	75
400	100

Lesson 29/pages 79-81

1. 259.2

259.2	24
1,080	100

2. 198

198	99
200	100

3. 330

330	75
440	100

4. 414

414	23
1,800	100

5. 630

630	10
6,300	100

6. 90

90	18
500	100

7. 13.75

13.75	5
275	100

8. 1

1	10
10	100

9. 255.2

255.2	58
440	100

10. 14

14	20
70	100

11. 64

64	80
80	100

12. 21

21	75
28	100

Lesson 30/pages 82-83

	Estimate	Actual
1.	$.70	$.68
2.	$2.31	$2.11
3.	$1.89	$1.76
4.	$7.70	$7.15
5.	$6.72	$6.24

Life Skill/page 84

1. $120
2. $150
3. $1.10
4. $1.50
5. $106.25
6. $478.80
7. $225

Lesson 31/pages 85-86

1. 248

248	32
775	100

2. 5,000

5,000	62.5
8,000	100

3. 76

76	38
200	100

4. $137.50

137.50	250
55	100

Lesson 32/pages 87-90

1. 25%

84	25
336	100

2. 20%

620	20
3,100	100

3. 25%

83	25
332	100

4. 25%

720	25
2,880	100

5. 100%

789	100
789	100

6. 15%

96	15
640	100

7. 60%

6	60
10	100

8. 55%

352	55
640	100

9. 100%

25	100
25	100

10. 25%

6	25
24	100

11. 20%

17	20
85	100

12. 10%

50	10
500	100

13. 400%

48	400
12	100

14. 50%

49	50
98	100

Lesson 33/pages 91-92

	Estimate	Actual
1.	20%	21%
2.	28%	30%
3.	34%	36%
4.	12%	13%

Life Skill/page 93

1. $3,786
2. 23%
3. 21%
4. $6\frac{1}{2}$%
5. 14%
6. $\frac{1}{2}$%
7. yes
8. 16%

Lesson 34/pages 94-95

1. $33\frac{1}{3}$%

240	$33\frac{1}{3}$
720	100

2. .8%

14.40	.8
1,800	100

3. $66\frac{2}{3}$%

34	$66\frac{2}{3}$
51	100

4. 4%

2,320	4
58,000	100

Lesson 35/pages 96-99

1. 6

3	50
6	100

2. 160

240	150
160	100

3. 4

6	150
4	100

4. 90

45	50
90	100

5. 800

440	55
800	100

6. 38

38	100
38	100

7. 1,055

884	80
1,055	100

8. 665

133	20
665	100

9. 800

528	66
800	100

10. 1,800

1,350	75
1,800	100

11. 200

2	1
200	100

12. 400

92	23
400	100

13. 700

77	11
700	100

14. 70

21	30
70	100

15. 250

28	8
250	100

16. 24

18	75
24	100

Lesson 36/pages 100-101

	Estimate	Actual
1.	$6,000	$5,591.25
2.	$4,000	$4,423.75
3.	$5,000	$5,027.75
4.	$7,000	$6,588.75

Life Skill/pages 102-103

1. $20.00
2. $22.50
3. $32
4. $50
5. $36
6. $36.80
7. $42
8. Answers will vary. Sample Answer: Best purchase can be based on quality, lowest price, best wear, or best savings.

Lesson 37/pages 104-105

1. $600

42	7
600	100

2. $55,600

278	.5
55,600	100

3. $45

6.75	15
45	100

4. $125,000

25,000	20
125,000	100

Lesson 38, pages 106-110

1. 45,000

45,000	$83\frac{1}{3}$
54,000	100

2. 1.1

1.1	$\frac{1}{4}$
440	100

3. 102

102	$66\frac{2}{3}$
153	100

4. 500

500	$166\frac{2}{3}$
300	100

5. 33

33	$33\frac{1}{3}$
99	100

6. 11.25

11.25	$7\frac{1}{2}$
150	100

7. $16\frac{2}{3}$%

16	$16\frac{2}{3}$
96	100

8. $83\frac{1}{3}$%

400	$88\frac{1}{3}$
480	100

9. $66\frac{2}{3}$%

50	$66\frac{2}{3}$
75	100

10. $12\frac{1}{2}$%

8	$12\frac{1}{2}$
64	100

11. $33\frac{1}{3}$%

1	$33\frac{1}{3}$
3	100

12. $83\frac{1}{3}$%

25	$83\frac{1}{3}$
30	100

13. 36

36	$16\frac{2}{3}$
216	100

14. 366

305	$83\frac{1}{3}$
366	100

15. 27

9	$33\frac{1}{3}$
27	100

16. 144

144	$37\frac{1}{2}$
384	100

17. 648

81	$12\frac{1}{2}$
648	100

18. 46

46	$66\frac{2}{3}$
69	100

19. $62\frac{1}{2}$%

40	$62\frac{1}{2}$
64	100

20. $33\frac{1}{3}$%

105	$33\frac{1}{3}$
315	100

21. 232

145	$62\frac{1}{2}$
232	100

22. 342

342	$66\frac{2}{3}$
513	100

Lesson 39/pages 111-115

1. $\frac{2}{9}$

$\frac{2}{9}$	50
$\frac{4}{9}$	100

2. $\frac{3}{4}$

$\frac{3}{4}$	150
$\frac{1}{2}$	100

3. $\frac{1}{3}$

$\frac{1}{3}$	40
$\frac{5}{6}$	100

4. $\frac{9}{50}$

$\frac{9}{50}$	20
$\frac{9}{10}$	100

5. $\frac{1}{3}$

$\frac{1}{3}$	50
$\frac{2}{3}$	100

6. $\frac{1}{2}$

$\frac{1}{2}$	75
$\frac{2}{3}$	100

7. 400%

$\frac{4}{5}$	400
$\frac{1}{5}$	100

8. 25%

$\frac{1}{20}$	25
$\frac{1}{5}$	100

9. 500%

$\frac{1}{2}$	500
$\frac{1}{10}$	100

10. 200%

$\frac{2}{5}$	200
$\frac{1}{5}$	100

11. 200%

$\frac{1}{2}$	200
$\frac{1}{4}$	100

12. 50%

$\frac{1}{3}$	50
$\frac{2}{3}$	100

13. $\frac{3}{4}$

$\frac{1}{2}$	$66\frac{2}{3}$
$\frac{3}{4}$	100

14. $1\frac{2}{3}$

$\frac{1}{2}$	30
$1\frac{2}{3}$	100

15. $\frac{4}{5}$

$\frac{1}{5}$	25
$\frac{4}{5}$	100

16. $\frac{8}{9}$

$\frac{2}{3}$	75
$\frac{8}{9}$	100

17. $\frac{1}{9}$

$\frac{1}{10}$	90
$\frac{1}{9}$	100

18. 1

$\frac{1}{3}$	$33\frac{1}{3}$
1	100

19. 50%

$\frac{1}{3}$	50
$\frac{2}{3}$	100

20. $\frac{4}{5}$

$\frac{1}{5}$	25
$\frac{4}{5}$	100

21. 20%

$\frac{1}{9}$	20
$\frac{5}{9}$	100

22. $\frac{1}{2}$

$\frac{1}{2}$	25
2	100

23. $\frac{3}{4}$

$\frac{3}{8}$	50
$\frac{3}{4}$	100

24. $33\frac{1}{3}\%$

$\frac{2}{7}$	$33\frac{1}{3}$
$\frac{6}{7}$	100

25. $\frac{5}{6}$

$\frac{1}{2}$	60
$\frac{5}{6}$	100

26. 125%

$\frac{2}{3}$	125
$\frac{8}{15}$	100

27. $\frac{1}{20}$

$\frac{1}{20}$	25
$\frac{1}{5}$	100

28. $\frac{1}{2}$

$\frac{3}{4}$	150
$\frac{1}{2}$	100

Lesson 40/pages 116-119

1. increase by 10's
2. increase by 4
3. 40
4. 4
5. 20
6. 32
7. Answer will vary. Sample Answer: When the percent increases by 10, the part increases by 4.
8. 24

9. 144
10. 48
11. 168
12. 72
13. 192
14. 96
15. 216
16. 120
17. 240
18. 15
19. 90
20. 30
21. 105
22. 45
23. 120
24. 60
25. 135
26. 75
27. 150
28. 55
29. 330
30. 110
31. 385
32. 165
33. 440
34. 220
35. 495
36. 275
37. 550

Lesson 41/pages 120-123

Answers will vary. Sample answers are given.
1. It is the value after the word *of*.
2. The percent has a percent sign.
3. (1) Seventy-five percent of four hundred is three hundred.
   (2) Three hundred is seventy-five percent of four hundred.
4. Multiply the diagonals of the two known values.
5. The divisor is in the diagonal with the unknown value.
6. The known diagonals can slant in either direction.
7. 65.6
8. 94
9. 120
10. 500
11. $33\frac{1}{3}\%$
12. $8\frac{1}{3}\%$
13. 510
14. 2,100
15. 582
16. 3
17. 50%
18. 170
19. 9
20. 25%

21. 75
22. 5,600
23. 100%
24. 20%
25. 10%
26. 240
27. 52.8
28. 5%
29. 300
30. 2,838

Lesson 42/pages 124-125

1. (2), (3)
2. (1), (3)
3. (1), (4)
4. (2), (3)
5. Multiplying by .01 is the same as dividing by 100 and multiplying a whole number by .01 is the same as moving the decimal point two places to the left.

Lesson 43/pages 126–127

1. 72
2. 600
3. 20%
4. 180
5. 90
6. 80%

Posttest/Unit 3/page 128

1. 550%
2. 3,000
3. 570
4. $66\frac{2}{3}\%$
5. 1,875
6. 0.375
7. $33\frac{1}{3}\%$
8. 400
9. 423
10. $33\frac{1}{3}\%$
11. $843.75
12. $60.75
13. $43.20
14. 4%
15. 25%

Pretest/Unit 4/pages 129-130

1. 10.5%
2. $1,000
3. $1,100
4. 10%
5. $96
6. 25%
7. 25%
8. $2.10
9. $45.50 $48.23
10. $300

Lesson 44/pages 131-132

1. 35%
2. 41%
3. 24%
4. 35%
5. 22%

Lesson 45/pages 133-135

1. (1) 35%
   (2) 6,000 bought souvenirs
   (3) 9,600

2. (1) 26%
   (2) 13 hours
   (3) 18 hours
3. (1) 20%
   (2) $60,000
   (3) $39,000
4. (1) 43%
   (2) 1.6 hours
   (3) 2.96 hours

Life Skill/page 136

Precinct 1
  Percent of Total, A   45%
  Percent of Total, B   55%
Precinct 2
  Votes for A   4,048
  Votes for B   3,952
Precinct 3
  Votes for A   2,500
  Percent of Total, B $66\frac{2}{3}\%$
Precinct 4
  Votes for A   6,975
  Percent of Total, B   38%
Precinct 5
  Votes for A   3,290
  Votes for B   6,110
Precinct 6
  Percent of Total, A   50%
  Votes for B   5,125
Grand Total
  Votes for B   28,037
  Percent of Total, B   53%
1. 52,900 votes
2. 28,037 votes
3. 47% of the total
4. 53% of the total
5. Candidate B

Lesson 46/pages 137-138

1. $125.00 each
2. 240 miles
3. 400 pounds
4. $120

Lesson 47/pages 139-140

1. 12.5%
2. 75%
3. 70%
4. 75%
5. 75%
6. $44\frac{4}{9}\%$

Life Skill/page 141

1. $15
2. $20
3. $96
4. $128
5. $4.80
6. $6.40
7. $4.80
8. $6.40

Lesson 48/pages 142-144

1. $333.90
2. $546.00
3. $97.50
4. $44.10
5. $835.80
6. $529.65

Lesson 49/pages 145-147

1. 20%
2. 20%
3. 15%
4. $33\frac{1}{3}$%
5. 22%
6. 40%

Life Skill/pages 148-149

1. 7.2
2. 1,448
3. 15%
4. 4,250
5. 61.74
6. 638.4
7. 8
8. 30%

Lesson 50/pages 150-151

1. 13,110
2. $562.50
3. $1,250
4. $250
5. $4.60
6. $2,750
7. $32,000
8. $90.

Life Skill/page 152

1. Washer
   Markup Percent 25%
   Selling Price $687.30
2. Dryer
   Markup Amount $181.30
   Selling Price $671.30
3. TV set
   Wholesale Price $1,000
   Selling Price $1,750
4. Freezer
   Markup Percent 30%
   Selling Price $897
5. Refrigerator
   Markup Amount $150
   Selling Price $600
6. Stereo
   Markup Amount $93.75
   Selling Price $468.75
7. TV set
   Markup Percent 50%
   Selling Price $1,200
8. Freezer
   Markup Amount $420
   Selling Price $1,120

Lesson 51/pages 153-155

1. 76%
2. 54 trucks
3. 10 hours
4. 828 people
5. $1.90

Life Skill/page 156

1. $1,764
2. $2,054.85
3. $1,740.48
4. $1,521.95
5. $1,537.35
6. $1,068.65
7. $2,646
8. $2,180.50

Posttest/Unit 4/pages 157-158

1. $24,600
2. $350
3. 6% = 144 people
4. 25%
5. 24 children's rides
6. 15%
7. $1.26
8. $2,220
9. $282
10. $64

Pretest/Unit 5/pages 159-160

1. $1,674
2. $1,012.50
3. $3,486
4. $626.40
5. $35.52
6. $75
7. $219.50
8. $33\frac{1}{3}$%
9. 14.25
10. $55.25
11. (1) $1,350
    (2) $22\frac{8}{11}$%
12. $165
13. $1,265
14. $85

Lesson 52/pages 161-162

1. (1) $80,000
   (2) $6%
   (3) $4,800
   (4) 1 year
2. (1) $5,000
   (2) 5%
   (3) $250
   (4) 1 year

3. (1) $750
   (2) 22%
   (3) $165
   (4) 1 year
4. (1) $1,000
   (2) 5%
   (3) $50
   (4) 1 year
5. (1) $3,000
   (2) 15%
   (3) $450
   (4) 1 year
6. (1) $1,500
   (2) 7%
   (3) $1,050
   (4) 10 years

Lesson 53/pages 163-164

1. $861
2. $3,240
3. $3,360
4. $33,800
5. $266.70
6. $108
7. $840
8. $688.50

Lesson 54/pages 165-166

1. (1), (2)
2. (1), (3)
3. (2), (4)
4. (1), (3)
5. (1), (2)

Lesson 55/pages 167-168

1. $13,230
2. $7,100
3. $23,800
4. $5,200
5. $17,500
6. $246,400
7. $10,140
8. $8,580

Lesson 56/pages 169-170

1. $114.80
2. $2,552.54
3. $434.60
4. $33.50
5. $2,772
6. $99.75
7. $826.20
8. $5,893.60
9. $36,309
10. $6,675

Lesson 57/pages 171-172

1.	$1,736	6.	$105.84
2.	$1,176	7.	$463.05
3.	$125	8.	$3,388
4.	$712.25	9.	$24.30
5.	$21.97	10.	$95.70

Lesson 58/page 173

1.	$10\frac{1}{4}$	8.	1
2.	$1\frac{1}{3}$	9.	$\frac{5}{6}$
3.	$6\frac{1}{2}$	10.	$4\frac{1}{2}$
4.	$\frac{1}{6}$	11.	$1\frac{1}{4}$
5.	$\frac{2}{3}$	12.	$5\frac{2}{3}$
6.	$\frac{1}{3}$	13.	$3\frac{2}{3}$
7.	$1\frac{1}{6}$	14.	$10\frac{3}{4}$

Life Skill/pages 174-175

1. $5,760
2. $5,760
3. $13,440
4. $4,838.40
5. $18,278.40
6. $380.80
7. $5,376
8. $18,816
9. $313.60
10. Federal Savings
11. Federal Savings

Lesson 59/pages 176-177

1.	$2,943	4.	$1,560
2.	$62.40	5.	$1,530
3.	$90	6.	$2,750

Lesson 60/pages 178-179

1.	$360	4.	$756
2.	$1,680	5.	$22
3.	$432	6.	$15.40

Life Skill/pages 180-181

1.	10	6.	$375
2.	2	7.	$835
3.	$535	8.	$6.35
4.	6	9.	$350
5.	$10.67	10.	$535.67

Lesson 61/page 182

1.	12	3.	1,095
2.	24	4.	20

Lesson 62/pages 183-185

1. (1) $320
   (2) $6,720
   (3) $336
   (4) $7,056
   (5) $640
   (6) Answers will vary. Sample Answer: The one using compound interest.
   (7) Answers will vary. Sample Answer: Because the interest earned during the first six months also earns interest during the second six months.
2. (1) $64
   (2) $2,624
   (3) $65.60
   (4) $2,689.60
   (5) $67.24
   (6) $2,756.84
   (7) $68.921 rounded to $68.92
   (8) $2,825.76

Life Skill/pages 186-187

1.	A	3.	A
2.	B	4.	B

Lesson 63/pages 188-189

1. (1) $59.85
   (2) $399.60
   (3) $459.45
   (4) $35.91
   (5) $24.54
2. (1) $268.75
   (2) $960
   (3) $1,228.75
   (4) $107.50
   (5) $46.25

Life Skill/pages 190-192

1. (1) 24%
   (2) $987.33
   (3) $54.77
   (4) $972.86
   (5) 2%
   (6) $19.75
2. (1) 21.6%
   (2) $76.50
   (3) $37.45
   (4) $38.83
   (5) 1.8%
   (6) $1.38

Lesson 64/pages 193-194

1. (1) $1.20 (2) $432.70
2. (1) $1.62 (2) $76.62
3. (1) $4.18 (2) $1,081.50
4. (1) $11.32 (2) $44.77

5. (1) $18 (2) $203.02
6. (1) $75.90 (2) $189.12

Lesson 65/pages 195-197

1. (1) $47.04
   (2) $558
   (3) $605.04
2. (1) $180
   (2) $10.80
   (3) $190.80
3. (1) $2,160
   (2) $8,000
   (3) 9%
   (4) 3 years

Posttest/Unit 5/pages 198-199

1. $14.73
2. $28
3. $18.75
4. $859.32
5. $151
6. $250
7. $2,700
8. $2,900
9. $200
10. Answers may vary. Sample answer: Installment plan because it would have the lowest monthly payments and the lowest cost—a savings of $200.

Unit 6/Pretest/page 200

1.	75	5.	(2)
2.	6	6.	(1)
3.	51	7.	(4)
4.	198	8.	(1)

Lesson 66/pages 201-202

1. 24 chairs to 6 tables
   24 chairs : 6 tables
   $\frac{24 \text{ chairs}}{6 \text{ tables}}$
2. 25 students : 1 teacher
   $\frac{25 \text{ students}}{1 \text{ teacher}}$
3. 2:1
4. 8:1
5. 1:10,000,000
6. 24:1
7. 1:25
8. 20:80 = 1:4
9. 25:1
10. 1:3
11. $\frac{2}{5}$
12. $\frac{125}{275} = \frac{5}{11}$
13. $\frac{3}{1}$
14. $\frac{5}{65} = \frac{1}{13}$

15. $\frac{24}{1,284} = \frac{2}{107}$

16. $\frac{336}{6} = \frac{56}{1}$

Lesson 67/pages 203-204

1. $\frac{1}{2}$

2. $\frac{1}{6}$

3. none shaded
4. none shaded

5. $\frac{3}{1}$

6. $\frac{1}{8}$

7. $\frac{2}{1}$

8. $\frac{1}{3}$

9. $\frac{3}{2}$

10. $\frac{2}{3}$

11. $\frac{1}{2}$

12. none shaded

13. $\frac{3}{5}$

14. none shaded

Lesson 68/pages 205-207

	Shaded	Unshaded	Total
Squares	2	2	4
Circles	4	2	6
Triangles	5	3	8
Diamonds	1	5	6
Total	12	12	24

1. $\frac{1}{6}$ 7. $\frac{4}{3}$

2. $\frac{1}{4}$ 8. $\frac{1}{1}$

3. $\frac{1}{6}$ 9. $\frac{1}{2}$

4. $\frac{2}{3}$ 10. $\frac{2}{1}$

5. $\frac{2}{3}$ 11. $\frac{3}{4}$

6. $\frac{1}{1}$ 12. $\frac{1}{1}$

Lesson 69/page 208

1. 12:18::44:66 or
   $\frac{12}{18} = \frac{44}{66}$

2. 27:54::45:90 or
   $\frac{27}{54} = \frac{45}{90}$

3. 9:81::15:135 or
   $\frac{9}{81} = \frac{15}{135}$

4. 10:100::100:1,000 or
   $\frac{10}{100} = \frac{100}{1,000}$

5. 150:300::75:150 or
   $\frac{150}{300} = \frac{75}{150}$

6. 99:1::990:10 or
   $\frac{99}{1} = \frac{990}{10}$

7. 348:58::162:27 or
   $\frac{348}{58} = \frac{162}{27}$

8. 17:51::1:3 or
   $\frac{17}{51} = \frac{1}{3}$

Lesson 70/pages 209-210

1. 216, 216
2. 105, 105
3. 55
4. 8
5. 3
6. 33
7. 42
8. 12
9. $\frac{3}{4}$
10. 7,744
11. $\frac{11}{18}$
12. 3
13. .2
14. 2.5

Life Skill/page 211

1. 1 inch
2. $\frac{3}{4}$ inch
3. 6 miles
4. $7\frac{1}{2}$ miles

Lesson 71/pages 212-213

1. 21 parks
2. 276 radio parts
3. 8.25 gallons
4. 3,262 miles

Life Skill/pages 214-215

1. $\frac{3}{8}$ lbs   11. $1\frac{1}{4}$ lbs

2. $\frac{1}{4}$ lbs   12. $\frac{5}{6}$ lbs

3. 3 tbs   13. 10 tbs

4. $2\frac{1}{4}$ oz   14. $7\frac{1}{2}$ oz

5. 6 oz   15. 20 oz

6. 12 oz   16. 40 oz

7. $\frac{3}{4}$ tsp   17. $2\frac{1}{2}$ tsp

8. $1\frac{1}{8}$ tsp   18. $3\frac{3}{4}$ tsp

9. $\frac{3}{8}$ tsp   19. $1\frac{1}{4}$ tsp

10. $1\frac{1}{2}$ tsp   20. 5 tsp

Life Skill/page 216

1. $8
2. $2.16
3. $6.21
4. $1.98
5. $3.28
6. $1.60
7. $4.92
8. $13.12

Posttest/Unit 6/page 217

1. 5
2. .1
3. 125
4. 106
5. $\frac{3}{2} = 1\frac{1}{2}$
6. 3
7. (1)
8. (4)
9. (2)
10. (3)

Practice for Mastering Percents and Proportions/pages 218-220

1. .15
2. $\frac{3}{20}$
3. 50%
4. $\frac{1}{2}$
5. 16.66
6. 39
7. 570
8. $66\frac{2}{3}$%
9. 141
10. 2,900 votes
11. 989 votes
12. $36
13. 4%
14. (1) 32%
    (2) 25%
    (3) $2,230
    (4) $301
    (5) Installment plan
    (6) $5,000
    (7) $2,400
    (8) 7,500 calls
15. (1) $1,125
    (2) $5,325
    (3) $810
    (4) $5,310
    (5) Installment plan
16. 50%
17. (1) 10%
    (2) $4,800